Compact Textbooks in Mathematics

Compact Textbooks in Mathematics

This textbook series presents concise introductions to current topics in mathematics and mainly addresses advanced undergraduates and master students. The concept is to offer small books covering subject matter equivalent to 2- or 3-hour lectures or seminars which are also suitable for self-study. The books provide students and teachers with new perspectives and novel approaches. They may feature examples and exercises to illustrate key concepts and applications of the theoretical contents. The series also includes textbooks specifically speaking to the needs of students from other disciplines such as physics, computer science, engineering, life sciences, finance.

- **compact:** small books presenting the relevant knowledge
- **learning made easy:** examples and exercises illustrate the application of the contents
- **useful for lecturers:** each title can serve as basis and guideline for a semester course/lecture/seminar of 2–3 hours per week.

More information about this series at http://www.springer.com/series/11225

Werner Ballmann

Introduction to Geometry and Topology

Birkhäuser

Werner Ballmann
Max-Planck-Institut für Mathematik
Bonn, Germany

Translated by Walker Stern, Germany

ISSN 2296-4568   ISSN 2296-455X   (electronic)
Compact Textbooks in Mathematics
ISBN 978-3-0348-0982-5   ISBN 978-3-0348-0983-2   (eBook)
https://doi.org/10.1007/978-3-0348-0983-2

Library of Congress Control Number: 2018942511

Printed on acid-free paper

This book is published under the imprint Birkhäuser, www.birkhauser-science.com by the registered company Springer International Publishing AG part of Springer Nature.
The registered company address is: Gewerbestrasse 11, 6330 Cham, Switzerland

FOR MY GRANDSONS, ALEX AND FINN.

Preface

The foundations of this book are the lecture notes for various courses which I summarized, revised, and expanded for an introductory course on geometry and topology. The text is conceived as the basis for a semester-long lecture course in the middle of a bachelor's program. The table of contents provides a good overview of the topics discussed.

I assume familiarity with linear algebra and real analysis of several variables. The first two chapters of the book are devoted to introductions to topological spaces and manifolds. Whether these concepts were already discussed in an analysis course depends on the objectives of the lecturer. If these concepts are not yet sufficiently familiar, one should begin with the first two chapters of the book. For a one-semester lecture course, one will then have to omit some material from the following chapters, as the text is likely too extensive for a one-semester course.

A problem in the modern curriculum is the fact that students must begin their bachelor theses at a time when they have not yet delved deeply enough into a subject area suitable for a thesis. I therefore attempt to give students experience with diverse topics, which they can then further explore in seminars and reading courses. At the end of each chapter, I have added references to supplementary literature intended as source material for possible student talks. Additionally, there are a great many textbooks on the topics treated in this text which I have not mentioned in the bibliography. Striving for completeness in this respect would have gone beyond the scope of this work.

Acknowledgements My thanks go to Bernd Ammann, Benedikt Fluhr, Karsten Große-Brauckmann, Hermann Karcher, Alexander Lytchak, Kaan Öcal, Anna Pratoussevitch, Dorothee Schueth, Juan Souto, Jan Swoboda, Thomas Vogel, and the many others who helped me improve my old notes and the text of the German edition with their many incisive recommendations. I am also grateful to Benedikt Fluhr for contributing the figures for that edition. My special thanks go to Walker Stern for translating the text into English and contributing new figures. We had many delightful discussions about how to transport meaning and humor from German into English, and I learned a lot from our conversations.

I would also like to thank the MPIM in Bonn for its support, and, in particular, for space and time. Last but not least, I would like to thank the staff of Birkhäuser Verlag for their help and cooperation throughout the publication process.

Bonn, Germany   Werner Ballmann

Contents

First Steps in Topology

Werner Ballmann

© Springer Basel 2018
W. Ballmann, *Introduction to Geometry and Topology*, Compact Textbooks in Mathematics,
https://doi.org/10.1007/978-3-0348-0983-2_1

In an analysis course, the reader will have already encountered metric spaces and terms like *open, closed, convergent, continuous,* and *compact.* These and other concepts are treated axiomatically in point-set topology.

In this chapter, we discuss the fundamentals of point-set topology. Here, as the propositions typically follow directly from the definitions, we will for the most part leave them as exercises for the reader. One of the exceptions is the Jordan Curve Theorem,[1] which we prove (following [CR]) for piecewise linear paths. After studying this chapter, the reader ought to be able to quickly and easily work through anything about point-set topology they are unfamiliar with. Good sources for this material are, for example, [Kel] and [La, Chapter I].

1.1 Topological Spaces

Definition 1.1.1

A *topology on a set* X is a subset $\mathcal{T}$ of the power set $\mathcal{P}(X)$ with the following properties:

1. $\emptyset \in \mathcal{T}$ and $X \in \mathcal{T}$;
2. Unions of elements of $\mathcal{T}$ belong to $\mathcal{T}$. That is, if $(U_i)_{i \in I}$ is a family of subsets of X, then

$$U_i \in \mathcal{T} \text{ for all } i \in I \implies \cup_{i \in I} U_i \in \mathcal{T};$$

3. Intersections of finitely many elements of $\mathcal{T}$ belong to $\mathcal{T}$. That is, if $(U_i)_{i \in I}$ is a finite family of subsets of X, then

---

(Continued)

---
[1] Marie Ennemond Camille Jordan (1838–1922).

Definition 1.1.1 (continued)

$$U_i \in \mathcal{T} \text{ for all } i \in I \Longrightarrow \cap_{i \in I} U_i \in \mathcal{T}.$$

A *topological space* is a set X together with a topology $\mathcal{T}$ on X. For a topological space $(X, \mathcal{T})$, we call the elements of $\mathcal{T}$ *open subsets* and their complements *closed subsets* of X.

It is a convention that the *empty union* of subsets of X is itself empty, and the *empty intersection* is equal to X. Therefore, if $I = \emptyset$ in condition (2) or (3), then $\bigcup_{i \in I} U_i := \emptyset$ and $\bigcap_{i \in I} U_i := X$. This sounds quite reasonable—as long as one can remember it. In any case, adopting this convention, condition (1) follows from conditions (2) and (3), and is in this sense superfluous. In the following, we will speak of a topological space X when it is either clear or irrelevant which topology on X we mean.

Example 1.1.2

1) Let X be a set. Then $\mathcal{T} = \{\emptyset, X\}$, is a topology on X, called the *trivial topology* (also sometimes referred to as the *indiscrete topology*). The only open sets of X in this topology are $\emptyset$ and X. It is not possible to have fewer open sets.

2) The power set $\mathcal{P}(X)$ of a set X is a topology on X, called the *discrete topology*. All subsets of X are open in this topology. It is not possible to have more open sets. A topological space is called *discrete* when its topology is discrete.

3) Call a subset U of $\mathbb{R}$ open if, for every $x \in U$, there is an $\varepsilon > 0$ with $(x - \varepsilon, x + \varepsilon) \subseteq U$. The set of such open subsets of $\mathbb{R}$ is a topology on $\mathbb{R}$, called the *canonical topology*.

4) Let X be a metric space, and let d denote the metric on X. Call a subset U of X open if, for every $x \in U$, there is an $\varepsilon > 0$ such that the open metric ball

$$B(x, \varepsilon) := \{y \in X \mid d(x, y) < \varepsilon\} \subseteq U.$$

The set of such open subsets of X is a topology $\mathcal{T}_d$ on X, called the *canonical topology* or the *topology associated to d*. A topological space $(X, \mathcal{T})$ is called *metrizable* if there is a metric d on X with $\mathcal{T} = \mathcal{T}_d$.

5) The set $\mathcal{T}_+ \subseteq \mathcal{P}(\mathbb{R})$ that consists of the subsets (a, ∞), $a \in [-\infty, \infty]$, is a topology on $\mathbb{R}$. Accordingly, one obtains a topology $\mathcal{T}_-$ on $\mathbb{R}$ with the subsets $(-\infty, b)$ for $b \in [-\infty, \infty]$. ∎

Definition 1.1.3

Let $\mathcal{T}$ be a topology on a set X. A subset $\mathcal{B} \subseteq \mathcal{T}$ is called a *basis* for $\mathcal{T}$ if every element of $\mathcal{T}$ is a union of elements of $\mathcal{B}$.

In Definition 1.1.3 we recall the convention that the empty union is empty. As a result, we have no need of more complicated formulations involving the empty set.

Proposition 1.1.4 *A subset $\mathcal{B}$ of a topology $\mathcal{T}$ on a set X is a basis of $\mathcal{T}$ if and only if, for every $U \in \mathcal{T}$ and $x \in U$, there is a $V \in \mathcal{B}$ with $x \in V \subseteq U$.* □

In formulating the following proposition, we again use the convention that the empty union is empty.

Proposition 1.1.5 *Let $\mathcal{B}$ be a subset of the power set $\mathcal{P}(X)$ of a set X with the following two properties:*
1. *X is the union of the elements of $\mathcal{B}$;*
2. *for each $B_1, B_2 \in \mathcal{B}$ and $x \in B_1 \cap B_2$ there is a $B_3 \in \mathcal{B}$ with $x \in B_3 \subseteq B_1 \cap B_2$.*
Let $\mathcal{T} \subseteq \mathcal{P}(X)$ be the subset whose elements are unions of elements of $\mathcal{B}$. Then $\mathcal{T}$ is a topology on X, and $\mathcal{B}$ is a basis of $\mathcal{T}$. □

Example 1.1.6
1) The set of open intervals (a, b) with $a, b \in \mathbb{Q}$ is a basis of the canonical topology on $\mathbb{R}$.
2) In a metric space, the set of open metric balls is a basis of the canonical topology. ∎

Proposition and Definition 1.1.7 *For $\mathcal{E} \subseteq \mathcal{P}(X)$, let $\mathcal{B} \subseteq \mathcal{P}(X)$ be the subset consisting of sets which are finite intersections of elements of $\mathcal{E}$. Then $\mathcal{B}$ satisfies the conditions of Proposition 1.1.5 and is therefore the basis of a topology, the topology generated by $\mathcal{E}$. We call $\mathcal{E}$ a* generating set *or a* sub-basis *of this topology.* □

Example 1.1.8
The canonical topology on $\mathbb{R}$ is generated by $\mathcal{T}_+ \cup \mathcal{T}_-$. Compare with Example 1.1.2.5. ∎

Definition 1.1.9

Let X be a topological space, $x \in X$ (resp. $Y \subseteq X$). Then $U \subseteq X$ is called a *neighborhood* of x (resp. Y) when there is an open set $x \in V \subseteq U$ (resp. $Y \subseteq V \subseteq U$). We denote by $\mathcal{U}(x)$ (resp. $\mathcal{U}(Y)$) the set of all neighborhoods of x (resp. Y).

Proposition 1.1.10 *A subset U of a topological space X is open if and only if U is a neighborhood of every point $x \in U$.* □

Definition 1.1.11

Let X be a topological space and $x \in X$. Then we call a subset $\mathcal{B}(x) \subseteq \mathcal{U}(x)$ a *neighborhood basis of* x if, for every neighborhood U of x, there is a $V \in \mathcal{B}(x)$ with $V \subseteq U$.

Example 1.1.12

Let X be a metric space and $x \in X$. Then the balls $B(x, 1/n)$, $n \in \mathbb{N}$, form a neighborhood basis of x. ∎

Definition 1.1.13

Let X be a topological space.
1. X satisfies the *first countability axiom* and is called *first countable* if every point in X admits a countable neighborhood basis.
2. X satisfies the *second countability axiom* and is called *second countable* if the topology of X admits a countable basis.

Example 1.1.14

1) All metric spaces satisfy the first countability axiom. Compare with Example 1.1.12.
2) The Euclidean space[2] $\mathbb{R}^n$ (with the canonical topology, i.e. the topology associated to the Euclidean metric) satisfies the second countability axiom. This is because the set of open balls with rational radii around points with rational coordinates is a countable basis of the topology. ∎

Definition 1.1.15

Let X be a topological space and $Y \subseteq X$. Then $x \in X$ is called
1. *an adherent point* (also sometimes called a *point of closure*) of Y, if every neighborhood of x in X contains a point of Y.[3] The set $\overline{Y}$ of limit points of Y is called the *closure* of Y;
2. *an interior point* of Y if there is a neighborhood of x in X that is contained in Y. The set $\mathring{Y}$ of interior points of Y is called the *interior* of Y;
3. *a boundary point* of Y if every neighborhood of x in X contains points of Y and $X \setminus Y$. The set of boundary points of Y is called the *boundary* of Y, here denoted by ∂Y.

---
[2] Euclid of Alexandria (ca. 360–280 BCE).

[3] A similar notion appearing commonly in the literature is that of *limit points* of Y. These are points $x \in X$ such that every neighborhood of x in X contains points of Y *other than x itself*. The notion of an *accumulation point* is also similar. It is important for the reader to observe that these notions are distinct.

Proposition 1.1.16 *Let X be a topological space and $Y \subseteq X$. Then the following hold:*

1. *$\overline{Y}$ is the smallest closed subset of X containing Y, and is therefore the intersection of all closed subsets of X containing Y.*
2. *$\mathring{Y}$ is the largest open subset of X contained in Y, and is therefore the union of all open subsets of X contained in Y.*
3. *$X \setminus \overline{Y} = \text{Interior}(X \setminus Y)$ and $\partial Y = \overline{Y} \setminus \mathring{Y}$. In summary, therefore, X is the disjoint union $X = \mathring{Y} \cup \partial Y \cup (X \setminus \overline{Y})$.* □

Definition 1.1.17

Let X be a topological space, and $Y \subseteq X$. Then Y is called
1. *dense* in X if $\overline{Y} = X$, and
2. *nowhere dense* in X if the interior of Y is empty.

Example 1.1.18
The set $\mathbb{Q}$ is dense in $\mathbb{R}$. The sets $Y := \{1/n \mid n \in \mathbb{N}\}$ and $\mathbb{Z}$ are nowhere dense in $\mathbb{R}$. ∎

1.2 Continuous Maps

Definition 1.2.1

Let X and Y be topological spaces and $f : X \longrightarrow Y$ be a function. We call f *continuous* if $f^{-1}(V)$ is open in X for all open V in Y. Or, equivalently, if $f^{-1}(A)$ is closed in X for all closed A in Y.

Proposition 1.2.2

1. *For every topological space X, id_X is continuous.*
2. *The composition of continuous maps is continuous.* □

Definition 1.2.3

Let X and Y be topological spaces and $f : X \longrightarrow Y$ a map. We call f *continuous at a point $x \in X$* if, for every neighborhood V of $f(x)$ in Y, there is a neighborhood U of x with $f(U) \subseteq V$.

Remark 1.2.4 For metric spaces this is equivalent to the usual $\varepsilon\delta$-definition.

Proposition 1.2.5 *Let X and Y be topological spaces and $f : X \longrightarrow Y$ a map. Then f is continuous if and only if f is continuous at all points $x \in X$.*   □

Proposition 1.2.6 *Let X and Y be topological spaces and $f : X \longrightarrow Y$ a map. Let $\mathcal{E}$ be a generating set of the topology of Y. Then f is continuous if and only if $f^{-1}(U)$ is open for all U in $\mathcal{E}$.*   □

Definition 1.2.7

A map $f : X \longrightarrow Y$ between topological spaces X and Y is called a *homeomorphism* if f is bijective and f and f^{-1} are continuous.

Example 1.2.8

The following maps are homeomorphisms:

$$\mathbb{R} \longrightarrow \mathbb{R}, x \mapsto x^3; \quad \mathbb{R} \longrightarrow (0, \infty), x \mapsto e^x; \quad (0, \infty) \longrightarrow (0, \infty), x \mapsto 1/x.$$

∎

Definition 1.2.9

We call a map $f : X \longrightarrow Y$ between topological spaces X and Y *open* if $f(U)$ is open in Y for every open U in X. We call f *closed* if $f(A)$ is closed in Y for every closed A in X.

Proposition 1.2.10 *Let X and Y be topological spaces and $f : X \longrightarrow Y$ a map. Then the following are equivalent:*

1. *f is a homeomorphism;*
2. *f is bijective, continuous, and open;*
3. *f is bijective, continuous, and closed.*   □

Definition 1.2.11

Let $\mathcal{T}_1$ and $\mathcal{T}_2$ be topologies on a set X. Then we call $\mathcal{T}_1$ *finer* than $\mathcal{T}_2$ and $\mathcal{T}_2$ *coarser* than $\mathcal{T}_1$ if $\mathcal{T}_1 \supseteq \mathcal{T}_2$.

The discrete topology is the finest possible topology, and the trivial topology is the coarsest possible.

Proposition 1.2.12 *Let $\mathcal{T}_1$ and $\mathcal{T}_2$ be topologies on a set X. Then the following are equivalent:*

1. *$\mathcal{T}_1$ is finer than $\mathcal{T}_2$;*
2. *the identity map* $\mathrm{id}\colon (X, \mathcal{T}_1) \longrightarrow (X, \mathcal{T}_2)$ *is continuous;*
3. *the identity map* $\mathrm{id}\colon (X, \mathcal{T}_2) \longrightarrow (X, \mathcal{T}_1)$ *is open.* □

Clearly, we have the following rule of thumb: a map $f\colon X \longrightarrow Y$ between topological spaces is more likely to be continuous the finer the topology on X or the coarser the topology on Y. For example, every such map is continuous if either X is equipped with the discrete topology or Y is equipped with the trivial topology.

1.3 Convergence and Hausdorff Spaces

Definition 1.3.1 ——————

Let X be a topological space and (x_n) a sequence in X. Then a point $x \in X$ is called a *limit* of the sequence (x_n) if, for every neighborhood U of x, there is an $n \in \mathbb{N}$ such that $x_m \in U$ for all $m \geq n$. We then say that the sequence *converges to x*, and we call the sequence *convergent*.

Remark 1.3.2 If X is a metric space, then this definition agrees with the one already familiar to the reader.

Proposition 1.3.3 *Let X and Y be topological spaces and $f\colon X \longrightarrow Y$ a map. Let $x \in X$ be a point with a countable neighborhood basis. Then f is continuous at x if and only if, for every sequence (x_n) with limit x in X, $f(x)$ is the limit of the sequence $f(x_n)$ in Y.* □

If the topology on X is trivial, then every sequence in X is convergent and every point of x is a limit of any sequence. It is therefore clear that the notion of convergence is not always sensible. We would like the limits of sequences to be unique, and so the *Hausdorff axiom*[4] enters the picture.

Definition 1.3.4 ——————

A topological space is called a *Hausdorff space* if, for every two points $x \neq y$ in X, there are neighborhoods U of x and V of y in X such that $U \cap V = \emptyset$.

Remark 1.3.5 The Hausdorff axiom is a so-called *separation axiom*, and is often referred to as T_2.

---
[4]Felix Hausdorff (1868–1942).

Example 1.3.6
Metric spaces are Hausdorff spaces.   ∎

Proposition 1.3.7 *Let X be a Hausdorff space. Then points $x \in X$ are closed subsets of X, and limits of sequences in X are unique (if they exist).*   □

If $x \in X$ is the unique limit of a sequence (x_n) in X, we write $\lim_{n \to \infty} x_n = x$, or, more concisely, $\lim x_n = x$.

1.4 New from Old

Proposition and Definition 1.4.1 *Let $(X, \mathcal{T})$ be a topological space, and $Y \subseteq X$. Then sets of the form $U = V \cap Y$ for $V \in \mathcal{T}$ constitute a topology on Y, called the* relative topology. *This topology is also known as the* subspace topology *or the* induced topology.   □

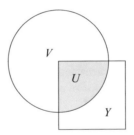

◘ Fig. 1.1 The relative topology

Example 1.4.2
Let X be a metric space with metric d_X, and let $Y \subseteq X$. Then Y together with the restriction d_Y of d_X to Y is a metric space. The metric balls around $x \in Y$ with respect to d_X and d_Y satisfy $B_Y(x, r) = B_X(x, r) \cap Y$. Therefore, the topology on Y associated to d_Y is precisely the relative topology with respect to the topology on X associated to d_X.   ∎

Proposition 1.4.3 *Let X be a topological space, and $Y \subseteq X$. Then:*
1. *The relative topology is the coarsest topology on Y such that the inclusion $i : Y \longrightarrow X$ is continuous.*
2. *If X is a Hausdorff space, then so is Y with the relative topology.*   □

Proposition 1.4.4 *Let X be a topological space and $Y \subseteq X$. Let $i : Y \longrightarrow X$ be the inclusion map. Then the relative topology on Y is characterized by the following so-called* universal property:
For all topological spaces Z and maps $f : Z \longrightarrow Y$, f is continuous if and only if $i \circ f$ is continuous.

Proof

The map $i : Y \longrightarrow X$ is continuous with respect to the relative topology. Therefore $i \circ f$ is continuous if f is continuous. On the other hand, let $i \circ f$ be continuous and $U \subseteq Y$ open in Y. Then there is a $V \subseteq X$, V open in X, such that $U = V \cap Y = i^{-1}(V)$, see 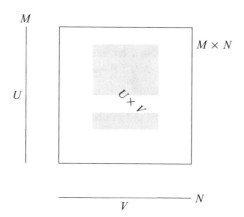 Fig. 1.1. Since $f^{-1}(U) = f^{-1}\left(i^{-1}(V)\right) = (i \circ f)^{-1}(V)$, it follows from the continuity of $i \circ f$ that $f^{-1}(U)$ is open in Z. Therefore it also follows that f is continuous, and the relative topology has the specified property.

Denote the topology on X by $\mathcal{T}$, and the relative topology on Y by $\mathcal{T}_1$. Let $\mathcal{T}_2$ be another topology on Y with the specified property. Since the identity map id : $(Y, \mathcal{T}_2) \longrightarrow (Y, \mathcal{T}_2)$ is continuous and $i = i \circ$ id, it follows that $i : (Y, \mathcal{T}_2) \longrightarrow (X, \mathcal{T})$ is continuous. Again from $i = i \circ$ id, it follows that id : $(Y, \mathcal{T}_2) \longrightarrow (Y, \mathcal{T}_1)$ is continuous. Analogously, one can show that id : $(Y, \mathcal{T}_1) \longrightarrow (Y, \mathcal{T}_2)$ is continuous. Therefore, the open sets of $\mathcal{T}_1$ and $\mathcal{T}_2$ are the same, and hence, $\mathcal{T}_1 = \mathcal{T}_2$. □

Proposition and Definition 1.4.5 *Let X and Y be topological spaces. Then the sets of the form $U \times V$, with U open in X and V open in Y form the basis of a topology on $X \times Y$. This topology is called the* product topology *(* *Fig. 1.2).* □

■ **Fig. 1.2** The product topology

Proposition 1.4.6 *Let X and Y be topological spaces. Then*
1. *The product topology is the coarsest topology on $X \times Y$ such that both projections $X \times Y \longrightarrow X$ and $X \times Y \longrightarrow Y$ are continuous.*
2. *If X and Y are Hausdorff, so is $X \times Y$ with the product topology.* □

The proof of the universal property of the product topology formulated in the proposition below is similar to the proof of the universal property of the relative topology given above. In Exercise 1.9.6 we will formulate a statement that includes both cases.

Proposition 1.4.7 *Let X and Y be topological spaces, and let $p_X : X \times Y \longrightarrow X$ and $p_Y : X \times Y \longrightarrow Y$ be the projections. Then the product topology is characterized by the following* universal property*:*
For all topological spaces Z and maps $f : Z \longrightarrow X \times Y$, f is continuous if and only if $p_X \circ f$ and $p_Y \circ f$ are continuous. $\qquad\square$

The definition and propositions carry over analogously to finite products of topological spaces. For arbitrary products $X = \prod_{i \in I} X_i$ of topological spaces, one defines the product topology as follows: A basis is given by sets of the form $U = \prod_{i \in I} U_i$, where U_i is open in X_i for all $i \in I$, and $U_i \neq X_i$ for at most finitely many $i \in I$. Analogues of Propositions 1.4.6 and 1.4.7 then hold. All of these notions fit into the general schema of the initial topology from Exercise 1.9.6.

Definition 1.4.8

Let X be a topological space, $R \subseteq X \times X$ an equivalence relation, and $\pi : X \longrightarrow Y$ the canonical projection from X to the set $Y := X/R$ of equivalence classes under R. Then the set $\mathcal{T}$ of subsets U of Y for which $\pi^{-1}(U)$ is open in X form a topology on Y, called the *quotient topology*.

Proposition 1.4.9 *The quotient topology $\mathcal{T}$ of Definition 1.4.8 has the following properties:*
1. *$\mathcal{T}$ is the finest topology on Y such that π is continuous.*
2. *For all topological spaces Z and maps $f : Y \longrightarrow Z$, f is continuous if and only if $f \circ \pi$ is continuous.* $\qquad\square$

Examples of the quotient topology will be discussed in Exercise 1.9.4. The quotient topology fits into the general schema of the final topology discussed in Exercise 1.9.7.

1.5   Connectedness and Path-Connectedness

Definition 1.5.1

A topological space X is called *connected* if there are no open subsets U and V of X such that $U \cup V = X$, $U \neq \emptyset$, $V \neq \emptyset$, and $U \cap V = \emptyset$. A subset Y of a topological space X is called connected if it is a connected topological space when equipped with the relative topology.

Remark 1.5.2
1) The empty set is connected.
2) In the definition, one can replace *open* with *closed*.
3) A subset Y of X is connected if and only if there are no open subsets U and V of X with $Y \subseteq U \cup V$, $U \cap Y \neq \emptyset$, $V \cap Y \neq \emptyset$, and $U \cap V \cap Y = \emptyset$.

Proposition 1.5.3 *A topological space is connected if and only if there are no non-empty proper subsets of X that are both open and closed.* □

Proposition 1.5.4 *For a topological space X, the following are equivalent:*
1. *X is connected.*
2. *There is no continuous, surjective map $f : X \longrightarrow \{0, 1\}$,*
   where $\{0, 1\}$ is equipped with the discrete topology.
3. *Continuous maps from X to discrete spaces are constant.*

Proof

If $f : X \longrightarrow \{0, 1\}$ is continuous and surjective, then $U = f^{-1}(0)$ and $V = f^{-1}(1)$ are non-empty open subsets of X with $U \cup V = X$ and $U \cap V = \emptyset$. If, on the other hand, U and V are non-empty, open subsets of X with $U \cup V = X$ and $U \cap V = \emptyset$, then $f : X \longrightarrow \{0, 1\}$ defined by $f(x) = 0$ for $x \in U$ and $f(x) = 1$ for $x \in V$ is a continuous surjective map with respect to the discrete topology on $\{0, 1\}$. The equivalence between (1) and (2) follows. The equivalence between (2) and (3) is left as an exercise for the reader. □

Proposition 1.5.5 *Let X and Y be topological spaces, and $f : X \longrightarrow Y$ a continuous map. Then if $Z \subseteq X$ is connected, $f(Z) \subseteq Y$ is connected.* □

Proposition 1.5.6 *Let X and Y be topological spaces. Then $X \times Y$ (equipped with the product topology) is connected if and only if X and Y are connected.*

Proof

Let p_X and p_Y be the projections from $X \times Y$ onto X and Y. These are continuous with respect to the given topologies. If $X \times Y$ is connected, then by Proposition 1.5.5, so are its images under p_X and p_Y, which are, respectively, X and Y.

On the other hand, let X and Y be connected. We assume that $X \times Y$ is not connected, and derive a contradiction. By Proposition 1.5.4 there is then a continuous, surjective map $f : X \times Y \longrightarrow \{0, 1\}$, and therefore points (x_0, y_0) and (x_1, y_1) in $X \times Y$ with $f(x_0, y_0) = 0$ and $f(x_1, y_1) = 1$. By the universal property of the product topology, the inclusions

$$i_0 : X \longrightarrow X \times Y, \quad i_0(x) := (x, y_0),$$

$$i_1 : Y \longrightarrow X \times Y, \quad i_1(y) := (x_1, y),$$

are continuous. Further, since X and Y are connected, Proposition 1.5.4.3 implies

$$(f \circ i_0)(x_0) = (f \circ i_0)(x_1) \quad \text{and} \quad (f \circ i_1)(y_0) = (f \circ i_1)(y_1).$$

Therefore, for the mixed point (x_1, y_0), we have

$$0 = f(x_0, y_0) = (f \circ i_0)(x_0) = (f \circ i_0)(x_1) = f(x_1, y_0)$$
$$= (f \circ i_1)(y_0) = (f \circ i_1)(y_1) = f(x_1, y_1) = 1,$$

which is a contradiction. Therefore $X \times Y$ is connected. □

Remark 1.5.7 An analogous form of the proposition holds for arbitrary products of topological spaces.

It follows from the Intermediate Value Theorem that a subset of the real numbers $\mathbb{R}$ is connected if and only if it is an interval (Exercise 1.9.8).

Definition 1.5.8

Let X be a topological space.

1. A *curve* or a *path* in X is a continuous map $c : I \longrightarrow X$, where I is an interval.
2. X is called *path-connected* if for every two points $x, y \in X$ there is a path $c : [a, b] \longrightarrow X$ with $c(a) = x$ and $c(b) = y$. In this case we say that c is *a path from x to y*. Furthermore, we call $Y \subseteq X$ *path-connected* if Y is path-connected with respect to the relative topology.

Remark 1.5.9 If $c : [a, b] \longrightarrow X$ is a path from x to y, then so is $c_1 : [0, 1] \longrightarrow X$, where $c_1(t) := c((1 - t)a + tb)$. In other words, we can restrict to the case of the unit interval in Definition 1.5.8.2.

Proposition 1.5.10 *If a topological space is path-connected, it is also connected.* □

Proposition 1.5.11 *Let X be a topological space, and $Y \subseteq X$. Then Y is path-connected if and only if, for each two points $x, y \in Y$, there is a path $c : [0, 1] \longrightarrow X$ from x to y whose image is contained in Y.* □

Proposition and Definition 1.5.12 *Let X be a topological space.*

1. If $c : [0, 1] \longrightarrow X$ is a path from x to y, the inverse path

$$c^{-1} : [0, 1] \longrightarrow X, \quad c^{-1}(t) := c(1 - t),$$

is a path from y to x.

2. If $c_0 : [0, 1] \longrightarrow X$ and $c_1 : [0, 1] \longrightarrow X$ are paths from x to y and y to z respectively, then the concatenation

$$c_0 * c_1 : [0, 1] \longrightarrow X, \quad (c_0 * c_1)(t) := \begin{cases} c_0(2t) & \text{if } 0 \leq t \leq 1/2, \\ c_1(2t - 1) & \text{if } 1/2 \leq t \leq 1, \end{cases}$$

is a path from x to z. □

Proposition and Definition 1.5.13 *Let X be a topological space. For $x \in X$ the* path component *of x in X is the set $P(x)$ of all points $y \in X$ such that there is a path from x to y. For all $x, y \in X$, the following hold:*

1. $x \in P(x)$;

2. $P(x)$ is path-connected;

3. $y \in P(x) \implies P(x) = P(y)$;
4. $y \in X \setminus P(x) \implies P(x) \cap P(y) = \emptyset$.

Therefore we obtain a decomposition of X into the distinct $P(x)$. We call these the path components *of X.*

Proof

(1) is clear. If $c_0 : [0, 1] \longrightarrow X$ and $c_1 : [0, 1] \longrightarrow X$ are paths from x to y and x to z respectively, then the concatenation $c_0^{-1} * c_1$ is a path from y to z, proving (2). The proof of (3) is similar, and (4) follows from (3). □

We obtain an analogous decomposition with connected subsets.

Lemma 1.5.14 *Let X be a topological space and $Y \subseteq X$. If Y is connected, then so are all $Y \subseteq Z \subseteq \overline{Y}$.*

Proof

Let U and V be open in X with $Z \subseteq U \cup V$ and $U \cap V = \emptyset$. Since Y is connected, and $Y \subseteq Z$, it follows that we can take $Y \subseteq U$ without loss of generality. Let $x \in Z$. Then x is an adherent point of Y, so every neighborhood of x contains points of Y. Either U or V is a neighborhood of x. From $Y \cap V = \emptyset$, it follows that $x \in U$, and therefore, $Z \subseteq U$. Therefore, Z is connected. □

Lemma 1.5.15 *Let X be a topological space, and (Y_i) a family of connected subsets of X. If $\bigcap_i Y_i \neq \emptyset$, then $\bigcup_i Y_i$ is connected.*

Proof

Let $x \in \bigcap_i Y_i$, and let U and V open in X with

$$\cup Y_i \subseteq U \cup V \quad \text{and} \quad U \cap (\cup Y_i) \cap V = \emptyset.$$

Then either $x \in U$ or $x \in V$; without loss of generality, we can take $x \in U$. For all $i \in I$, $U \cap Y_i$ and $V \cap Y_i$ are open in Y_i and

$$Y_i \subseteq U \cup V \quad \text{and} \quad U \cap Y_i \cap V = \emptyset.$$

Since $x \in U \cap Y_i$ and Y_i is connected, then $Y_i \subseteq U$ for all i. Therefore, $\bigcup_i Y_i$ is connected. □

Proposition and Definition 1.5.16 *Let X be a topological space. For $x \in X$ the* connected component *of x in X is the union $C(x)$ of all connected subsets of X containing x. The following hold for all $x, y \in X$:*

1. $x \in C(x)$;
2. $C(x)$ *is closed and connected;*
3. $y \in C(x) \implies C(x) = C(y)$;
4. $y \in X \setminus C(x) \implies C(x) \cap C(y) = \emptyset$.

Therefore, we obtain a decomposition of X into the distinct C(x). We call these the connected components *of X.*

Proof

(1) is clear. From Lemmas 1.5.14 and 1.5.15 it follows that $C(x)$ is closed and connected. If $y \in C(x)$, then there is a connected subset of X that contains x and y. Then, however, we also have $x \in C(y)$. From (2) it then easily follows that $C(x) \subseteq C(y)$ and $C(y) \subseteq C(x)$, proving (3) and (4).   □

Remark 1.5.17 For a topological space X and a point $x \in X$, $P(x)$ is always a subset of $C(x)$, because $P(x)$ is path connected.

Remark 1.5.18 If X has only finitely many connected components, then they are all open in X. On the other hand, for instance, we have $\mathbb{Q}$ with the induced topology from $\mathbb{R}$, which is *totally disconnected* in the sense that $C(x) = \{x\}$ for all $x \in \mathbb{Q}$. In particular, the connected components of $\mathbb{Q}$ are not open.

Definition 1.5.19

A topological space X is called *locally connected* (resp. *locally path-connected*) if, for every point $x \in X$ and every neighborhood U of x in X, there is a connected (resp. path-connected) neighborhood V of x in X with $V \subseteq U$.

Remark and Example 1.5.20

1) Open subsets of locally connected topological spaces are locally connected. Open subsets of locally path-connected topological spaces are locally path-connected.
2) Locally path-connected spaces are locally connected.
3) The Euclidean space $\mathbb{R}^k$ is locally path-connected.

Proposition 1.5.21 *For a topological space X, the following hold:*

1. *If X is locally connected, the connected components of X are open in X.*
2. *If X is locally path connected, then the path components of X are open in X. In particular, the path-components of X agree with the connected components of X.*   □

1.6   Compact Spaces

Definition 1.6.1

A family $(U_i)_{i \in I}$ of subsets of a set X is called a *cover* of a subset $Y \subseteq X$ if $Y \subseteq \bigcup_{i \in I} U_i$. We call a cover $(U_i)_{i \in I}$ of $Y \subseteq X$ *finite* if I is finite. If X is a topological space, we call a cover $(U_i)_{i \in I}$ of $Y \subseteq X$ *open* if U_i is open in X for all $i \in I$.

Definition 1.6.2

A topological space X is called *compact* if every open cover $(U_i)_{i \in I}$ of X contains a finite subcover,

$$X = U_{i_1} \cup \cdots \cup U_{i_k} \quad \text{with } i_1, \ldots, i_k \in I.$$

A subset Y of a topological space is called *compact* if it is compact with respect to the relative topology; in other words, Y is compact if every cover of Y by open subsets of X contains a finite subcover.

Proposition 1.6.3 *A topological space X is compact if and only if the following holds: A family $(A_i)_{i \in I}$ of closed subsets of X has non-empty intersection if all finite subfamilies of $(A_i)_{i \in I}$ have non-empty intersection. (Recall that $\bigcap_{i \in \emptyset} A_i = X$.)* □

Proposition 1.6.4 *A closed subset of a compact space is compact. A compact subset of a Hausdorff space is closed.*

Proof

Let X be a compact space, and $A \subseteq X$ a closed subset. Let $(U_i)_{i \in I}$ be an open cover of A. Then $(U_i)_{i \in I}$ together with $X \setminus A$ is an open cover of X. Since X is compact, there is a finite subset $J \subseteq I$ such that X is covered by $(U_i)_{i \in J}$ and $X \setminus A$. Then $(U_i)_{i \in J}$ is a finite cover of A.

Let X be a Hausdorff space, and $B \subseteq X$ a compact subset. Let $x \in X \setminus B$. Since X is a Hausdorff space, for every $y \in B$ there are open neighborhoods U_y of x and V_y of y such that $U_y \cap V_y = \emptyset$. Then $(V_y)_{y \in B}$ is an open cover of B. Since B is compact, this cover contains a finite subcover of B. In other words, there are points $y_1, \ldots, y_n$ in B with

$$B \subseteq V_{y_1} \cup \cdots \cup V_{y_n} =: V_x.$$

Then, however, V_x and

$$U_x := U_{y_1} \cap \cdots \cap U_{y_n}$$

are disjoint open neighborhoods of B and x respectively. In particular, U_x is a neighborhood of x contained in $X \setminus B$. Therefore, $X \setminus B$ is open, and B is closed. □

Proposition 1.6.5 *Let X be a Hausdorff space and $A, B \subseteq X$ be compact subsets with $A \cap B = \emptyset$. Then there are open neighborhoods U of A and V of B in X with $U \cap V = \emptyset$.*

Proof

From the proof of Proposition 1.6.4, there are, for every $x \in A$, disjoint open neighborhoods U_x of x and V_x of B. Since A is compact, there are points $x_1, \ldots, x_m$ in A such that

$$A \subseteq U_{x_1} \cup \cdots \cup U_{x_m} =: U.$$

Then, however, U and

$$V := V_{x_1} \cap \cdots \cap V_{x_m}$$

are disjoint open neighborhoods of A and B. □

Definition 1.6.6

Let X be a topological space. We call $x \in X$ an *accumulation point* of a family $(x_i)_{i \in I}$ of points in X if every neighborhood of x contains infinitely many members of the family $(x_i)_{i \in I}$. That is, for every neighborhood U of x, the set of $i \in I$ such that $x_i \in U$ is infinite.

Proposition 1.6.7 *If X is a compact topological space, then every infinite family $(y_i)_{i \in I}$ of points in X has an accumulation point.*

Proof

If this is not the case, then every $x \in X$ has an open neighborhood U_x containing at most finitely many elements of the sequence. That is, there are only finitely many $i \in I$ with $y_i \in U_x$. The family $(U_x)_{x \in X}$ covers X. Since X is compact, there are points $x_1, \ldots, x_n$ with

$$X = U_{x_1} \cup \cdots \cup U_{x_n}.$$

This, however, implies $|I| < \infty = |I|$, which is a contradiction. □

Definition 1.6.8

A topological space is said to be *sequentially compact* if every sequence in X has a subsequence which converges in X.

Proposition 1.6.9 *For a subset K of a metric space, the following are equivalent:*
1. *K is compact.*
2. *K is sequentially compact.*
3. *K is complete, and for every $\varepsilon > 0$ there are points $x_1, \ldots, x_n \in K$ with*

$$K \subseteq B(x_1, \varepsilon) \cup \cdots \cup B(x_n, \varepsilon).$$

Proof

(1) $\Rightarrow$ (2) follows from Proposition 1.6.7 together with Exercise 1.9.13.

(2) $\Rightarrow$ (3): The completeness of K is a direct consequence of (2). Let $\varepsilon > 0$ be given, and suppose K is not contained in a finite union of metric balls of radius ε as in (3). Let $x_1 \in K$. By assumption, there is then $x_2 \in K \setminus B(x_1, \varepsilon)$. By iterating this procedure, we obtain a sequence (x_n) in K with

$$x_{n+1} \in K \setminus B(x_1, \varepsilon) \cup \cdots \cup B(x_n, \varepsilon).$$

By (2), this sequence has a subsequence that converges in K. Let x be the limit of this subsequence. Then there are $m > n \geq 1$ with $d(x, x_m) < \varepsilon/2$ and $d(x, x_n) \leq \varepsilon/2$, contradicting $x_m \in K \setminus B(x_n, \varepsilon)$.

(3) $\Rightarrow$ (1): Let $(U_i)_{i \in I}$ be an open cover of K which contains no finite subcover of K. By (3), for a chosen $\alpha \in (0, 1)$, there is a finite cover of K by metric balls $B(x, \alpha)$. Therefore, there is a point $x_1 \in K$ such that $B(x_1, \alpha) \cap K$ cannot be covered by finitely many of the U_i. As in the case of K, $B(x_1, \alpha \cap K)$ can also be covered by finitely many metric balls $B(x, \alpha^2)$ with $x \in K$. Therefore, there is a point $x_2 \in K$ with

$$B(x_1, \alpha) \cap K \cap B(x_2, \alpha^2) \neq \emptyset$$

and such that $B(x_2, \alpha^2) \cap K$ cannot be covered by finitely many of the U_i. Iterating this procedure, we obtain a sequence (x_n) in K with

$$B(x_{n-1}, \alpha^{n-1}) \cap K \cap B(x_n, \alpha^n) \neq \emptyset$$

and such that $B(x_n, \alpha^n) \cap K$ cannot be covered by finitely many of the U_i. From this construction we obtain that $d(x_n, x_{n+1}) \leq 2\alpha^n$, and thus the sequence (x_n) is a Cauchy sequence.[5] Since K is complete, this sequence converges in K. The limit $x \in K$ of this sequence is contained in one of the sets U_i of the cover. Since U_i is open, there is an $\varepsilon > 0$ such that $B(x, \varepsilon) \subseteq U_i$. Then, however, $B(x_n, \alpha^n) \subseteq U_i$ for all sufficiently large n, which is a contradiction.

□

Corollary 1.6.10 (Heine-Borel Theorem[6]) *A subset of $\mathbb{R}^n$ is compact if and only if it is closed and bounded.*

□

Proposition 1.6.11 *The image of a compact subset under a continuous map is compact.* □

Corollary 1.6.12 *Let X be compact, and $f : X \longrightarrow \mathbb{R}$ continuous. Then f has a maximum.*

□

---

[5] Augustin-Louis Cauchy (1789–1857).

[6] Heinrich Eduard Heine (1821–1881), Félix Édouard Justin Émile Borel (1871–1956).

Proposition 1.6.13 *Let X be compact, Y a Hausdorff space, and $f : X \longrightarrow Y$ a continuous map. Then f is closed. If f is injective, then f is a homeomorphism onto its image.*   □

Definition 1.6.14

We call a topological space X *locally compact* if every point in X has a neighborhood basis consisting of compact subsets of X.

1.7   The Jordan Curve Theorem

A *Jordan curve* is a curve $c : [a, b] \longrightarrow \mathbb{R}^2$ such that $c|_{[a,b)}$ is injective and $c(a) = c(b)$. The first example that comes to mind is the circle

$$c \colon [0, 2\pi] \longrightarrow \mathbb{R}^2, \quad c(t) = (\cos t, \sin t).$$

The complement of the circle has two connected components,

$$B = \{x \in \mathbb{R}^2 \mid \|x\| < 1\} \quad \text{and} \quad A = \{x \in \mathbb{R}^2 \mid \|x\| > 1\},$$

with B bounded and A unbounded. The Jordan Curve Theorem says that an analogous property holds for all Jordan curves.

Jordan Curve Theorem 1.7.1 *Let $C \subseteq \mathbb{R}^2$ be the image of a Jordan curve c. Then $\mathbb{R}^2 \setminus C$ has two connected components, one of which is bounded and the other unbounded.*

We will not prove the Jordan Curve Theorem in full generality here, as this would be quite technically demanding. A short, though not particularly instructive, proof is possible with the appropriate tools from singular homology (see, for example, [Ha, section 2B.1]). The methods of this homological proof are quite similar to those used in the proof of the Jordan-Brouwer Separation Theorem 3.6.7 presented in Chap. 3. Indeed, a special case of the Jordan Curve Theorem follows directly from the Jordan-Brouwer Separation Theorem. Here, however, we aim for an elementary and instructive proof, and so restrict ourselves to the case in which c is a *piecewise linear curve*. That is, there is a subdivision

$$a = t_0 < t_1 < \cdots < t_k = b,$$

such that, for all $1 \leq i \leq k$ and $t \in [t_{i-1}, t_i]$,

$$c(t) = \frac{t_i - t}{t_i - t_{i-1}} c(t_{i-1}) + \frac{t - t_{i-1}}{t_i - t_{i-1}} c(t_i).$$

To simplify the presentation, we take $[a, b] = [0, 1]$ and extend the curve periodically to $\mathbb{R}$. The fact that c is a Jordan curve is then expressed as

$$c(t) = c(t') \iff t - t' \in \mathbb{Z}.$$

In other words, c is injective modulo $\mathbb{Z}$. The periodic extension of c to $\mathbb{R}$ also gives rise to an extension of the subdivision of $[0, 1]$, which we will number by $\mathbb{Z}$ according to $t_{jk+i} = t_i + j$ for integers $0 \le i \le k$ and j. After these preparations, we come to the proof of Theorem 1.7.1 for piecewise linear Jordan curves. The idea for the proof is taken from [CR].

Proof
The goal of the first stage of the proof is the construction of parallel, piecewise linear Jordan curves $c_s : \mathbb{R} \longrightarrow \mathbb{R}^2$ from $c = c_0$ whose images are pairwise disjoint. To this end, for $i \in \mathbb{Z}$ we set (see ◲ Fig. 1.3)

$$e_i := \frac{c(t_i) - c(t_{i-1})}{\|c(t_i) - c(t_{i-1})\|} =: (x_i, y_i) \quad \text{and} \quad f_i := (-y_i, x_i).$$

Then (e_i, f_i) is a positively oriented orthonormal basis of $\mathbb{R}^2$ (with the canonical orientation). Let ϕ_i be the oriented angle between $c(t_{i+1}) - c(t_i)$ and $c(t_i) - c(t_{i-1})$, so that

$$c(t_{i+1}) - c(t_i) = \|c(t_{i+1}) - c(t_i)\| \big(\cos(\phi_i) e_i + \sin(\phi_i) f_i \big).$$

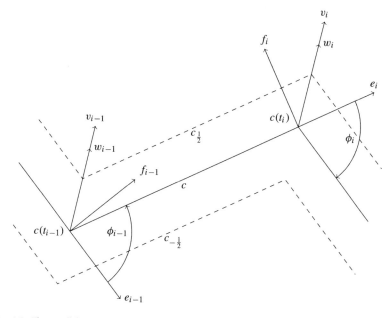

◲ **Fig. 1.3** The parallel piecewise linear curves

The angle bisector between $c(t_{i-1}) - c(t_i)$ and $c(t_{i+1}) - c(t_i)$ (and thus between $-e_i$ and e_{i+1}) points in the direction

$$w_i := \cos\left(\frac{\pi + \phi_i}{2}\right) e_i + \sin\left(\frac{\pi + \phi_i}{2}\right) f_i.$$

We now set

$$v_i := \frac{1}{\cos(\phi_i/2)} w_i.$$

For all s with $|s|$ sufficiently small, $c(t_{i-1}) + s v_{i-1} \neq c(t_i) + s v_i$, and the line through these points is then parallel to the line through $c(t_{i-1})$ and $c(t_i)$ at a distance $|s|$. From this we obtain a family of piecewise linear curves $c_s : \mathbb{R} \longrightarrow \mathbb{R}^2$ with

$$c_s(t) := \frac{t_i - t}{t_i - t_{i-1}}(c(t_{i-1}) + s v_{i-1}) + \frac{t - t_{i-1}}{t_i - t_{i-1}}(c(t_i) + s v_i)$$

for $i \in \mathbb{Z}$ and $t \in [t_{i-1}, t_i]$. By definition, $c_0 = c$. Since c is a Jordan curve, there is an $\varepsilon > 0$ such that

$$(-\varepsilon, \varepsilon) \times [t_{i-1} - \varepsilon, t_i + \varepsilon] \longrightarrow \mathbb{R}^2, \quad (s, t) \mapsto c_s(t),$$

is injective for all $i \in \mathbb{Z}$. Since c is injective modulo $\mathbb{Z}$, it follows easily by *reductio ad absurdum* that

$$(-\varepsilon, \varepsilon) \times \mathbb{R} \longrightarrow \mathbb{R}^2, \quad (s, t) \mapsto c_s(t),$$

is also injective modulo $\mathbb{Z}$ for ε chosen to be sufficiently small. Therefore, the c_s defined this way are piecewise linear Jordan curves and have pairwise disjoint images. This concludes the first stage of the proof.

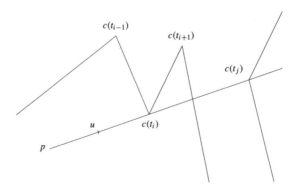

☐ Fig. 1.4 The intersection at $c(t_i)$ is not counted, the one at $c(t_j)$ is.

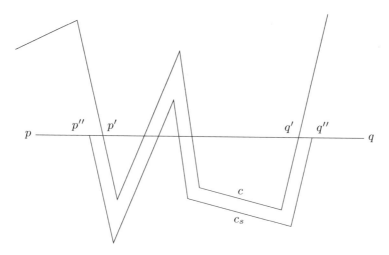

■ **Fig. 1.5** Going from p to q in A.

Let C be the image of c, and $u \neq 0$ a vector in $\mathbb{R}^2$ which is not a multiple of one of the vectors $c(t_i) - c(t_{i-1})$. We now define two sets $A, B \subseteq \mathbb{R}^2 \setminus C$ as follows: $p \in \mathbb{R}^2 \setminus C$ belongs to A (resp. B) when the number of intersections of the ray $\{p + tu \,|\, t \geq 0\}$ with C is even (resp. odd). In this definition, we count intersections with corners $c(t_i)$ as follows: if the piecewise linear curve $c|_{[t_{i-1}, t_{i+1}]}$ lies on one side of the ray, we do not count the intersection (we count it as 0), otherwise, we count it as a real intersection (as 1), see ■ Fig. 1.4.

By definition, $A \cap B = \emptyset$ and $A \cup B = \mathbb{R}^2 \setminus C$. If one side of a segment of c belongs to A (locally), then the other belongs to B, and vice versa. In particular, A and B are non-empty. Next we see that A and B are open. Therefore, paths in $\mathbb{R} \setminus C$ remain either in A or in B. So a path running from A into B must intersect C. Thus, the various c_s remain either entirely in A or in B, and c_{-s} remains in B when c_s remains in A (and vice versa).

With these insights, we arrive at the main point, namely, that A and B are path-connected. To this end, let $p, q \in A$. Since A is open, we may assume that the segment from p to q does not contain any of the points $c(t_i)$. If the segment from p to q has no intersection with C, there is nothing to prove. Otherwise, there is a first intersection point with C, $p' = c(t_1')$, and a last intersection point with C, $q' = c(t_2') \neq p'$, see ■ Fig. 1.5. For p'' on the segment between p and p' and sufficiently near to p', there is then a small $s \neq 0$ and t_1'' near to t_1' such that $p'' = c_s(t_1'')$. There is then also a t_2'' near t_2', such that $c_s(t_2'') = q''$ lies on the segment from p to q. We now claim that q'' lies between q' and q. Otherwise, $q'' \in B$ would contradict the fact that c_s remains entirely in A. Iterating this argument, we then see that p, p'', q'', and q all lie in one path-component of $\mathbb{R}^2 \setminus C$. Therefore, A is path-connected. Similarly, one finds that B is path-connected.

B is bounded, because when C lies in a disc of radius R, then the boundary of this disc belongs to A.

$\qquad\qquad\qquad\qquad\qquad\qquad\qquad\qquad\qquad\qquad\qquad\qquad\qquad\qquad\qquad\qquad\quad$ $\square$

A generalization of the Jordan Curve Theorem is due to Schoenflies.[7]

Proposition 1.7.2 (Schoenflies Theorem) *Let $C \subseteq \mathbb{R}^2$ be the image of a Jordan curve. Then there is a homeomorphism of $\mathbb{R}^2$ that sends C to the unit circle.*

While the analogue of the Jordan Curve Theorem holds in higher dimensions, the higher-dimensional version of the theorem of Schoenflies does not. In dimension three, *Alexander's horned sphere*[8] is a counter-example. On Wikipedia one can find a detailed discussion of the theorems of Jordan and Schoenflies, along with many references. A veritable treasure trove on this topic is the website of Andrew Ranicki.[9]

1.8   Supplementary Literature

The sources [Kel] and [La, Chapter I] mentioned at the beginning of this chapter are quite readable, even when only reading individual sections, and more than adequately cover the typical requirements of point-set topology.

1.9   Exercises

Exercise 1.9.1

1) Let $f : X \longrightarrow Y$ be a map between topological spaces. Then f is continuous if and only if, for all subsets B of Y, $f^{-1}(\mathring{B})$ is contained in the interior of $f^{-1}(B)$. Formulate a corresponding statement for the closure.
2) Let $\mathcal{T}$ be the canonical topology on $\mathbb{R}$ and $\mathcal{S} = \mathcal{T}_{\pm}$ be as in Example 1.1.2.5. Identify the set of all continuous maps $(\mathbb{R}, \mathcal{T}) \longrightarrow (\mathbb{R}, \mathcal{S})$.
3) Let $I, J \subseteq \mathbb{R}$ be intervals and $f : I \longrightarrow J$ a function. Then f is a homeomorphism if and only if f is strictly monotone and surjective.

■

Exercise 1.9.2

Let X be a topological space and $A \subseteq X$. If there is a sequence (x_n) in A that converges to $x \in X$, then $x \in \overline{A}$. If X is first countable, then the converse holds.   ■

Exercise 1.9.3

1) A topological space X is a Hausdorff space if and only if the diagonal $\{(x, x) \mid x \in X\}$ is closed in $X \times X$ with respect to the product topology.
2) Let X and Y be metric spaces with metrics d_X and d_Y. Then

$$d_\infty((x_1, y_1), (x_2, y_2)) := \max\{d_X(x_1, x_2), d_Y(y_1, y_2)\}$$

---

[7] Arthur Moritz Schoenflies (1853–1928).
[8] James Waddell Alexander II (1888–1971).
[9] Andrew Alexander Ranicki (1948–2018).

is a metric on $X \times Y$. Show that the topology associated to d_∞ on $X \times Y$ is the product topology of the topologies on X and Y associated to d_X and d_Y respectively. Also verify that the topology associated to the metric

$$d_s((x_1, y_1), (x_2, y_2)) := (d_X(x_1, x_2)^s + d_Y(y_1, y_2)^s)^{1/s}, \quad 1 \le s < \infty,$$

on $X \times Y$ agrees with that associated to d_∞.

■

Exercise 1.9.4

1) (Gluing the ends of a rope) Let $I = [0, 1]$ and $R = \{(x, x) \mid x \in I\} \cup \{(0, 1), (1, 0)\}$. Show that I/R, equipped with the quotient topology, is homeomorphic to the circle $S^1 := \{(y, z) \in \mathbb{R}^2 \mid y^2 + z^2 = 1\}$.
2) (Winding up a rope) Let R be the equivalence relation on $\mathbb{R}$ with $x \sim y$ if $x - y \in \mathbb{Z}$. Show that $\mathbb{R}/R$, equipped with the quotient topology, is homeomorphic to the circle S^1.
3) Analogously, let R be the equivalence relation on $\mathbb{R}^m$ with $x \sim y$ if $x - y \in \mathbb{Z}^m$. Show that $\mathbb{R}^m/R$, equipped with the quotient topology, is homeomorphic to the *torus* $T^m := (S^1)^m$ equipped with the product topology.

■

Exercise 1.9.5

Let the topological space X be the union of finitely many closed subsets X_α, each of which is equipped with the relative topology. Show that a subset of X is open if and only if its intersections with all of the X_α is open. Conclude that a map from X to a topological space Y is continuous if and only if its restriction to X_α is continuous for every α. ■

Exercise 1.9.6 (Initial Topology)

Let Y be a set, $(X_i, \mathcal{T}_i)$, $i \in I$, a family of topological spaces, and $g_i : Y \longrightarrow X_i$, $i \in I$, a family of maps. Show that the *initial topology* $\mathcal{T}$ on Y, generated by the sets $g_i^{-1}(V)$, $i \in I$ and $V \in \mathcal{T}_i$, is characterized by each of the two following properties:
1. $\mathcal{T}$ is the coarsest topology on Y, such that all the g_i are continuous.
2. For all topological spaces Z and maps $f : Z \longrightarrow Y$, f is continuous if and only if all of the $g_i \circ f$ are continuous.

Also verify that the relative and product topologies fit into this schema. Under which conditions will $(Y, \mathcal{T})$ be a Hausdorff space? ■

Exercise 1.9.7 (Final Topology)

Let Y be a set, $(X_i, \mathcal{T}_i)$, $i \in I$, a family of topological spaces, and $g_i : X_i \longrightarrow Y$, $i \in I$, a family of maps. Show that the set $\mathcal{T}$ of subsets U of Y, such that $g_i^{-1}(U)$ is open in X_i

for all $i \in I$, defines a topology on Y, the so-called *final topology*, and that this topology is characterized by each of the following properties:

1. $\mathcal{T}$ is the finest topology on Y, such that all of the g_i are continuous.
2. For all topological spaces Z and maps $f : Y \longrightarrow Z$, f is continuous if and only if all the $f \circ g_i$ are continuous.

Verify that the quotient topology fits into this schema. ∎

Exercise 1.9.8
A subset of the real numbers $\mathbb{R}$ is connected if and only if it is an interval. (Hint: Intermediate Value Theorem) ∎

Exercise 1.9.9
1) An open subset of $\mathbb{R}^m$ is path-connected if and only if it is connected.
2) The unit sphere $S^m = \{x \in \mathbb{R}^{m+1} \mid ||x|| = 1\}$ is path-connected for all $m \geq 1$.
3) The subset $\{(x, \sin(1/x)) \mid x > 0\} \cup \{(0, y) \mid y \in \mathbb{R}\}$ of $\mathbb{R}^2$ is connected, but not path-connected.
∎

Exercise 1.9.10
The subset of $\mathbb{R}^2$, comprised of points (x, y) with $x = 0$ or $y = 0$ or $y = 1/n$ with $n \in \mathbb{N}$ is path-connected with respect to the relative topology, but not locally connected. ∎

Exercise 1.9.11
Let $A, B \subseteq X$ be closed subsets. Show that A and B are connected if $A \cap B$ and $A \cup B$ are connected. ∎

Exercise 1.9.12
The graph of a continuous function $f : X \longrightarrow Y$ is path-connected if and only if X is path-connected. ∎

Exercise 1.9.13
Let X be a topological space, and $x \in X$ a point with a countable neighborhood basis. Then x is an accumulation point of the sequence (x_n) in X if and only if a subsequence of (x_n) converges to x. ∎

Exercise 1.9.14
A second-countable Hausdorff space is compact if and only if it is sequentially compact. ∎

Exercise 1.9.15 (The *Cantor Discontinuum*[10] or *Cantor Set*)

Let $C_0 := [0, 1]$. Remove the open middle third $(1/3, 2/3)$ of C_0, to obtain $C_1 := [0, 1/3] \cup [2/3, 1]$. Remove each of the open middle thirds of the sub-intervals of C_1, to obtain

$$C_2 := [0, 1/9] \cup [2/9, 1/3] \cup [2/3, 7/9] \cup [8/9, 1].$$

Recursively, remove each of the open middle thirds of the sub-intervals of C_n, $n \geq 2$, to get a decreasing sequence

$$C_0 \supseteq C_1 \supseteq C_2 \supseteq C_3 \supseteq \cdots$$

of closed subsets of $\mathbb{R}$. The *Cantor discontinuum* is the subset $C := \cap_{n=0}^{\infty} C_n \subseteq \mathbb{R}$. Verify that C is compact, uncountable, and nowhere dense in $\mathbb{R}$. ∎

Exercise 1.9.16

1) A Hausdorff space X is locally compact if and only if every point of X has a compact neighborhood.
2) A subset of a locally compact Hausdorff space X is closed if and only if its intersection with every compact subset of X is closed. ∎

Exercise 1.9.17

Let $(X, \mathcal{T})$ be a locally compact Hausdorff space which is not compact. Define a new topological space whose underlying set is

$$X^* := X \sqcup \{\infty\},$$

where ∞ denotes a single point which we append to X. The topology $\mathcal{T}^*$ on X^* is defined to be

$$\mathcal{T}^* := \mathcal{T} \cup \{U = (X \setminus C) \cup \{\infty\} \mid C \text{ is compact in } X\}.$$

The topological space $(X^*, \mathcal{T}^*)$ is called the *one-point compactification of X*.
1. Show that $(X^*, \mathcal{T}^*)$ is compact.
2. Show that the map

$$f : X \longrightarrow X^*, \quad f(x) = x,$$

is a homeomorphism onto its image. Note that the complement of $f(X)$ in X^* consists of the single point ∞.
3. Prove that if $X = \mathbb{R}^m$, there is a homeomorphism $g : X^* \longrightarrow S^m$. (Hint: see Example 2.1.2.2.) Remember this fact when we come to the proof of the Jordan-Brouwer Separation Theorem 3.6.7. ∎

---

[10]Georg Ferdinand Ludwig Philipp Cantor (1845–1918).

Manifolds

Werner Ballmann

© Springer Basel 2018
W. Ballmann, *Introduction to Geometry and Topology*, Compact Textbooks in Mathematics,
https://doi.org/10.1007/978-3-0348-0983-2_2

For many problems both in mathematics and beyond, manifolds are the natural class of underlying spaces with which to work. From the perspective of analysis, manifolds are locally indistinguishable from Euclidean spaces, and are therefore tailor-made for use with the tools of analysis. Manifolds provide the natural setting for many concepts from analysis.

2.1 Manifolds and Smooth Maps

The key to the definition of manifolds is the chain rule: if U, V and W are open subsets of Euclidean space, and $f : U \longrightarrow V$ and $g : V \longrightarrow W$ are differentiable, then $g \circ f$ is differentiable, and the derivatives satisfy

$$D(g \circ f)|_x = Dg|_{f(x)} \circ Df|_x \quad \text{for all } x \in U.$$

It is also important that differentiability is a local property.

2.1.1 Atlases

For open subsets $U \subseteq \mathbb{R}^m$ and $V \subseteq \mathbb{R}^n$, we say that $f : U \longrightarrow V$ is a C^k map if f is continuous ($k = 0$), f is k-times continuously differentiable ($k \in \mathbb{N}$), f is infinitely differentiable ($k = \infty$), or f is real analytic[1] ($k = \omega$). Typically, we say *smooth* instead of *infinitely differentiable*. Each of these regularity classes is stable under restriction and composition, and these are the crucial properties for the following definition and discussion.

---

[1] Real analytic means that the components of f can be locally expressed as convergent power series. The reader who is unfamiliar with real analytic maps may replace *real analytic* with *infinitely differentiable*.

Definition 2.1.1

For $m \geq 0$ and $k \in \{0, 1, 2, \ldots, \infty, \omega\}$, an *m-dimensional C^k atlas* $\mathcal{A}$ on a set M is comprised of:

1. a cover $(U_i)_{i \in I}$ of M,
2. a family $(U_i')_{i \in I}$ of open subsets of $\mathbb{R}^m$, and
3. a family $x_i : U_i \longrightarrow U_i', i \in I$, of bijections,

such that $x_i(U_i \cap U_j)$ is open in $\mathbb{R}^m$ and the

$$x_i \circ x_j^{-1} : x_j(U_i \cap U_j) \longrightarrow x_i(U_i \cap U_j)$$

are C^k maps for all $i, j \in I$. See ◻ Fig. 2.1. We call the x_i the *charts*, the U_i the *coordinate neighborhoods*, and the $x_i \circ x_j^{-1}$ the *transition functions* of the atlas. We will also speak of *coordinates* or *local coordinates* rather than of charts.

The transition functions are invertible, since $(x_i \circ x_j^{-1})^{-1} = x_j \circ x_i^{-1}$. Therefore, the transition functions are homeomorphisms, or diffeomorphisms for the appropriate differentiability class. We will also call a C^0 atlas a *topological atlas*, a C^∞ atlas a *smooth atlas*, and a C^ω atlas a *real analytic atlas*. By definition, real analytic atlases are smooth, and for all $k \in \mathbb{N}$ smooth atlases are C^k and C^k atlases are C^{k-1}.

An atlas $\mathcal{A}$ on a set M is determined by the family of pairs (U_i, x_i), each of which consists of a coordinate neighborhood and a chart, and so we will often refer to such families as atlases.

Example 2.1.2

1) $(\mathbb{R}^m, \mathrm{id})$ is a real analytic atlas on $\mathbb{R}^m$ (with one chart).
2) Let $S^m = \{x \in \mathbb{R}^{m+1} \mid \|x\| = 1\}$ be the unit sphere in $\mathbb{R}^{m+1}$. The two subsets

$$U_+ = \{x \in S^m \mid x^0 \neq 1\} \quad \text{and} \quad U_- = \{x \in S^m \mid x^0 \neq -1\}$$

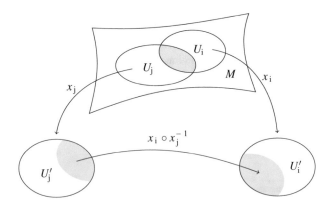

◻ **Fig. 2.1** Transition functions

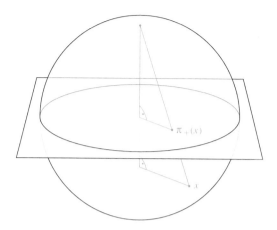

◘ Fig. 2.2 The stereographic projection π_+

cover S^m. The *stereographic projections*

$$\pi_\pm : U_\pm \longrightarrow \mathbb{R}^m, \quad \pi_\pm(x) := \frac{1}{1 \mp x^0}(x^1, \ldots, x^m),$$

each determine the intersection point $\pi_\pm(x)$ of the line through the north (or south) pole and x with $\mathbb{R}^m \cong \{x^0 = 0\} \subseteq \mathbb{R}^{m+1}$ (see ◘ Fig. 2.2) and are bijective.[2] The transition functions simplify to

$$(\pi_+ \circ \pi_-^{-1})(x) = (\pi_- \circ \pi_+^{-1})(x) = x/\|x\|^2.$$

Therefore $\mathcal{A} = ((U_+, \pi_+), (U_-, \pi_-))$ is a real analytic atlas on S^m.

■

For an atlas $\mathcal{A}$ on a set M there is a topology $\mathcal{T}_\mathcal{A}$ on M canonically associated to $\mathcal{A}$. In this way, we view a set M together with an atlas as a topological space:

Proposition 2.1.3 *Let M be a set and $\mathcal{A} = ((U_i, x_i))_{i \in I}$ an atlas on M. Let $\mathcal{T}_\mathcal{A}$ be the set of subsets U of M, such that $x_i(U \cap U_i)$ is open in $\mathbb{R}^m$ for all $i \in I$. Then $\mathcal{T}_\mathcal{A}$ is a topology on M, for which the following properties hold:*

1. *The subsets U of M for which there is an i with $U \subseteq U_i$, such that $x_i(U)$ is open in $\mathbb{R}^m$ form a basis of $\mathcal{T}_\mathcal{A}$.*
2. *The $x_i : U_i \longrightarrow U_i'$ are homeomorphisms.*
3. *Together with $\mathcal{T}_\mathcal{A}$, M is a locally compact, locally path-connected topological space.* □

Remark 2.1.4

1) $\mathcal{T}_\mathcal{A}$ is the final topology on M induced by the maps $x_i^{-1} : U_i' \longrightarrow M$ in the sense of Exercise 1.9.7

---

[2]In this example, x is not a chart, but rather the name of the variables.

2) In the Examples 2.1.2.1 and 2.1.2.2, the topologies induced by the respective atlases agree with the usual topologies on those sets.

Proposition and Definition 2.1.5 *Let $\mathcal{A} = ((U_i, x_i))_{i \in I}$ be an m-dimensional C^k atlas on a set M. Let $U \subseteq M$, $U' \subseteq \mathbb{R}^m$ be open subsets and $x : U \longrightarrow U'$ be a bijection. Then we call (U, x) a* chart compatible with $\mathcal{A}$, *if $x_i(U \cap U_i)$ and $x(U \cap U_i)$ are open in $\mathbb{R}^m$ and the*

$$x \circ x_i^{-1} : x_i(U \cap U_i) \longrightarrow x(U \cap U_i)$$

and their inverses

$$x_i \circ x^{-1} : x(U \cap U_i) \longrightarrow x_i(U \cap U_i)$$

are C^k maps for all $i \in I$.

The family $\overline{\mathcal{A}}$ of all charts compatible with an m-dimensional C^k atlas $\mathcal{A}$ on M is an m-dimensional C^k atlas on M, and is maximal *in the sense that it is not strictly contained in a larger m-dimensional C^k atlas on M. Such a maximal atlas is also called an m-*dimensional C^k structure *on M. In the case $k = \infty$ or $k = \omega$ one also speaks of* smooth *or* real analytic structures *respectively.*   □

Example 2.1.6
Let $U_i^{\pm} = \{x \in S^m \mid \pm x^i > 0\}$ and

$$\pi_i^{\pm} : U_i^{\pm} \longrightarrow \{u \in \mathbb{R}^m \mid \|u\| < 1\}, \quad \pi_i^{\pm}(x) := (x^0, \ldots, \hat{x}^i, \ldots, x^m),$$

where the hat indicates, that we omit x^i. Then the $(U_i^{\pm}, \pi_i^{\pm})$ form a real analytic atlas on S^m, the charts of which are compatible with the atlas from Example 2.1.2.2.   ■

Proposition and Definition 2.1.7 *Let $\mathcal{A}$ and $\mathcal{B}$ be atlases on a set M. If all the charts from $\mathcal{A}$ are compatible with $\mathcal{B}$, then*

$$\mathcal{T}_{\mathcal{A}} = \mathcal{T}_{\mathcal{B}}, \quad \overline{\mathcal{A}} = \overline{\mathcal{B}},$$

and we call $\mathcal{A}$ and $\mathcal{B}$ equivalent.   □

To work effectively, it is important to choose the best possible atlases, that is, atlases with the fewest possible charts and the simplest possible transition functions—structures remain in the background. This is one interpretation of Propositions 2.1.5 and 2.1.7. In this sense, a manifold is, in essence, a set M together with an atlas $\mathcal{A}$. However, some such pairs are true monsters,[3] and we therefore require two more properties of the topology $\mathcal{T}_{\mathcal{A}}$ induced by $\mathcal{A}$ to exclude unpleasant examples. For a moment, we set aside

---

[3] A term borrowed from Imre Lakatos (1922–1974).

the definition of a manifold and first discuss these two conditions, namely, that $\mathcal{T}_A$ be a Hausdorff topology and paracompact.

Example 2.1.8 (The Line with Two Origins)

Let $M := (-\infty, 0) \cup \{+i, -i\} \cup (0, \infty)$. We choose the two subsets

$$U_\pm = (-\infty, 0) \cup \{\pm i\} \cup (0, \infty)$$

as coordinate neighborhoods on M which cover M, and we choose as charts the maps

$$\kappa_\pm \colon U_\pm \longrightarrow \mathbb{R}, \quad \begin{cases} \kappa_\pm(x) = x & \text{for } x \neq \pm i, \\ \kappa_\pm(x) = 0 & \text{for } x = \pm i. \end{cases}$$

Then $\kappa_+(U_+ \cap U_-) = \kappa_-(U_+ \cap U_-) = \mathbb{R} \setminus \{0\}$, and the transition functions are

$$\kappa_+ \circ \kappa_-^{-1} = \kappa_- \circ \kappa_+^{-1} = \mathrm{id}.$$

The atlas $\mathcal{A} = ((U_+, \kappa_+), (U_-, \kappa_-))$ is therefore real analytic. The set M together with the topology $\mathcal{T}_A$ is certainly not a Hausdorff space, as we have doubled the origin (and, for pedagogical reasons, called the two copies of the origin $\pm i$). The existence of an atlas $\mathcal{A}$ on a set M therefore in no way guarantees that $\mathcal{T}_A$ is a Hausdorff topology. ∎

2.1.2 Paracompact Spaces

We next discuss the concept of paracompactness for general topological spaces, which we will therefore again denote by X in this interlude.

Definition 2.1.9

1. A cover $(V_j)_{j \in J}$ of a set X is called *finer* than a cover $(U_i)_{i \in I}$ of X if, for every V_j there exists a U_i with $V_j \subseteq U_i$.
2. A cover $(U_i)_{i \in I}$ of a topological space X is called *locally finite*, if every point of X has a neighborhood, that intersects only finitely many of the U_i.
3. A topological space X is called *paracompact*, if every open cover of X has a locally finite refinement.

Warning: The term *finer* in Definition 2.1.9,1 should be treated with caution. For example, every cover of a set X is finer than the cover by the whole power set.

Topological spaces which are not paracompact are not geometrical objects, as the following theorem demonstrates.

Proposition 2.1.10 *Metric spaces are paracompact.*

We will not use Proposition 2.1.10, and therefore will also not prove it; for a proof in greater generality see, for example, [Kel, Chapter 5].

Definition 2.1.11

An *exhaustion by compact sets* of a locally compact Hausdorff space X consists of a sequence (K_n) of compact subsets of X with

$$K_n \subseteq \overset{\circ}{K}_{n+1} \quad \text{and} \quad \cup K_n = X.$$

Proposition 2.1.12 *Let X be a locally compact Hausdorff space.*
1. *If the topology of X has a countable basis, then X has an exhaustion by compact sets.*
2. *If X is paracompact and connected, then X has an exhaustion by compact sets.*
3. *If X has an exhaustion by compact sets, then X is paracompact.*

Proof
1) Choose a countable basis (B_n) of the topology on X, such that the $\overline{B}_n$ are compact. (There is such a basis!) We recursively define a sequence of compact subsets K_n of X: Set $K_1 := \overline{B}_1$. Now suppose $K_1, \ldots, K_n$ have already been defined. Choose the smallest $m > n$ such that

$$K_n \subseteq B_1 \cup \cdots \cup B_m$$

and set

$$K_{n+1} := \overline{B}_1 \cup \cdots \cup \overline{B}_m.$$

Then for the sequence of the K_n,

$$K_n \subseteq \overset{\circ}{K}_{n+1} \quad \text{and} \quad \cup_n K_n = X.$$

In other words, X has a exhaustion by compact sets.
2) Choose a locally finite open cover $(U_i)_{i \in I}$ of X, such that the $\overline{U}_i$ are compact. Then every compact subset of X intersects only finitely many of the U_i.

Choose an $i = i(1) \in I$ with $U_{i(1)} \neq \emptyset$ and set $K_1 := \overline{U}_{i(1)}$. Then K_1 is compact and therefore only intersects finitely many of the U_i. We number these,

$$K_1 \cap U_{i(j)} \neq \emptyset, \quad i(1), \ldots, i(j_2) \in I,$$

and set

$$K_2 := \overline{U}_{i(1)} \cup \cdots \cup \overline{U}_{i(j_2)}.$$

Then K_2 is compact and therefore only intersects finitely many of the U_i. We number these in a way consistent with the numbering of the U_i in the previous step,

$$K_2 \cap U_{i(j)} \neq \emptyset, \quad i(1), \dots, i(j_3) \in I,$$

and set

$$K_3 := \overline{U}_{i(1)} \cup \cdots \cup \overline{U}_{i(j_3)}.$$

Recursively, we obtain a sequence of compact subsets K_n such that

$$K_n \subseteq U_{i(1)} \cup \cdots \cup U_{i(n+1)} = \mathring{K}_{n+1};$$

in particular $\cup_n K_n$ is open in X. Now let x be an adherent point of $\cup_n K_n$. A compact neighborhood of x in X only intersects finitely many of the $U_{i(j)}$, so x lies in the finite union of the closures of these $U_{i(j)}$. This union is contained in one of the K_n, and therefore so is x. Thus $\cup_n K_n$ is closed. Since X is connected and $\cup_n K_n$ is non-empty, it therefore follows that $X = \cup_n K_n$. Therefore (K_n) is an exhaustion of X by compact sets.

3) Now let $(K_n)_{n \geq 1}$ be an exhaustion by compact sets, and $\mathcal{U} = (U_i)_{i \in I}$ be an open cover of X. Choose finite open covers $(V_{nj})_{1 \leq j \leq k_n}$ of the compact sets $K_n \setminus \mathring{K}_{n-1}$, such that all V_{nj} are contained in $\mathring{K}_{n+1} \setminus K_{n-2}$ (with $K_{-1} = K_0 := \emptyset$) and such that for every V_{nj} there exists a U_i with $V_{nj} \subseteq U_i$. Then the cover of X by the V_{nj} is open, locally finite, and a refinement of the cover $\mathcal{U}$. $\qquad\square$

Sets with atlases $\mathcal{A}$, whose induced topologies $\mathcal{T}_A$ are Hausdorff but not paracompact, to wit, the *long line* and the *Prüfer surface*,[4] are discussed in detail in Appendix A1 of [Sp1].

Proposition 2.1.13 *Let $\mathcal{A} = ((U_i, x_i))_{i \in I}$ be an atlas on a set M, possessing both of the following properties:*
1. For all $p \neq q$ in M there are $i, j \in I$ and subsets $V_i \subseteq U_i$ and $V_j \subseteq U_j$ with

$$p \in V_i, \quad q \in V_j \text{ and } V_i \cap V_j = \emptyset,$$

such that $x_i(V_i)$ and $x_j(V_j)$ are open in $\mathbb{R}^m$,
2. I is at most countable.
Then M, together with the topology $\mathcal{T}_A$ induced by $\mathcal{A}$, is a locally compact Hausdorff space with a countable basis for its topology. In particular, M equipped with $\mathcal{T}_A$ is paracompact. $\qquad\square$

---

[4]Ernst Paul Heinz Prüfer (1896–1934).

2.1.3 Manifolds

Definition 2.1.14

An m-dimensional C^k manifold is a set M together with an m-dimensional C^k structure $\mathcal{A}$ on M such that M together with the topology $\mathcal{T}_\mathcal{A}$ induced by $\mathcal{A}$ is a paracompact Hausdorff space.

We will always view a manifold M as being equipped with the topology induced by its structure or by an atlas. Since, by Propositions 2.1.5 and 2.1.7, a structure consists of all charts that are compatible with an atlas contained in it, it is sufficient, in any given example, to specify an atlas. We then call the compatible charts the *charts of M*. For a point $p \in M$, a *chart around* p is a chart (U, x) on M with $p \in U$.

C^0 manifolds are also called *topological*, C^∞ manifolds *smooth* and C^ω manifolds *real analytic manifolds*. Connected manifolds of dimension 1 are also called *curves*. Examples of curves are the line $\mathbb{R}$ and the circle S^1. Manifolds of dimension 2 are also called *surfaces*. First examples are the plane $\mathbb{R}^2$ and the sphere S^2.

We now introduce a small selection of examples, which we will discuss repeatedly from a variety of viewpoints.

Example 2.1.15
1) *Vector spaces:* $\mathbb{R}^m$ with the atlas $(\mathbb{R}^m, \mathrm{id})$ is a real analytic manifold. More generally, let V be an m-dimensional vector space over $\mathbb{R}$. To a basis $B = (b_1, \ldots, b_m)$ of V, we associate a bijection

$$\iota_B : \mathbb{R}^m \longrightarrow V, \quad \iota_B(u) := u^i b_i,$$

where we make use of the Einstein summation convention.[5],[6] We then set $x_B := \iota_B^{-1}$ and thereby obtain an atlas $\mathcal{A} = ((V, x_B))_B$ on V, the transition functions of which are linear maps, and therefore real analytic. The topology induced by $\mathcal{A}$ is the canonical topology. This is a second countable and locally compact Hausdorff topology. Therefore V, together with the real analytic structure induced by $\mathcal{A}$, is a connected real analytic manifold of dimension m.
2) *Open subsets:* An open subset W of an m-dimensional C^k manifold M, together with the charts (U, x) of M with $U \subseteq W$, is itself canonically a C^k manifold of dimension m.
3) *Spheres:* The unit sphere S^m with the atlas from Example 2.1.2.2 is a compact real analytic manifold of dimension m.

---
[5] The *Einstein summation convention* stipulates that we sum over any indices that occur both as subscripts and as superscripts.
[6] Albert Einstein (1879–1955).

4) *Projective spaces* $\mathbb{K}P^n$ with $\mathbb{K} \in \{\mathbb{R}, \mathbb{C}, \mathbb{H}\}$[7,8]: By definition, $\mathbb{K}P^n$ is the set of one-dimensional $\mathbb{K}$-linear subspaces of $\mathbb{K}^{n+1}$. A point L in $\mathbb{K}P^n$ is determined by its *homogeneous coordinates*, $L = [x^0, \ldots, x^n]$, where $(x^0, \ldots, x^n)$ is a vector in $L \setminus \{0\}$. The $n + 1$ sets

$$U_i = \{[x^0, \ldots, x^n] \in \mathbb{K}P^n \mid x^i \neq 0\}, \quad 0 \leq i \leq n,$$

cover $\mathbb{K}P^n$, and we can define bijections/charts

$$\kappa_i : U_i \longrightarrow \mathbb{K}^n, \quad \kappa_i([x]) := (x^0, \ldots, \hat{x}^i, \ldots, x^n)(x^i)^{-1},$$

where the hat over x^i indicates that x^i is omitted. The transition functions are given by

$$(\kappa_i \circ \kappa_j^{-1})(x^1, \ldots, x^n) = (x^1, \ldots, \hat{x}^i, \ldots, 1, \ldots, x^n)(x^i)^{-1},$$

where the 1 is in the j-th place (and the formula represents the case $i < j$). Therefore $\mathcal{A} = ((U_i, \kappa_i))_{0 \leq i \leq n}$ is a dn-dimensional real analytic atlas on $\mathbb{K}P^n$ with $d = \dim_{\mathbb{R}} \mathbb{K} \in \{1, 2, 4\}$. Since $\mathcal{A}$ satisfies both conditions from Proposition 2.1.13, $\mathbb{K}P^n$ together with $\mathcal{A}$ is a dn-dimensional real analytic manifold. Later we will see that $\mathbb{K}P^n$ is compact.

5) *Grassmannian manifolds*[9] $G_k(V)$: For a vector space V of dimension n over $\mathbb{K} \in \{\mathbb{R}, \mathbb{C}, \mathbb{H}\}$ and a number $k \in \{1, \ldots, n - 1\}$, we denote by $G_k(V)$ the set of all k-dimensional $\mathbb{K}$-linear subspaces of V. This example thus generalizes the previous one. For a basis $E = (e_1, \ldots, e_n)$ of V, let $U_E \subset G_k(V)$ be the set of all subspaces of V, that can be written as the graph of a linear map

$$\mathrm{Span}(e_1, \ldots, e_k) := V_E \longrightarrow W_E := \mathrm{Span}(e_{k+1}, \ldots, e_n).$$

In other words, $P \in G_k(V)$ is in U_E if and only if there is a matrix $(a_\nu^\mu) \in \mathbb{K}^{(n-k) \times k}$ such that the tuple of vectors

$$e_\nu + \sum_{k < \mu \leq n} e_\mu a_\nu^\mu, \quad 1 \leq \nu \leq k,$$

is a basis of P. Then we set

$$\kappa_E(P) := (a_\nu^\mu) \in \mathbb{K}^{(n-k) \times k}$$

and thereby obtain a bijection/chart $\kappa_E : U_E \longrightarrow \mathbb{K}^{(n-k) \times k}$.

---

[7] We denote by $\mathbb{H}$ the space of Hamilton's quaternions. We set the convention that vectors in $\mathbb{H}^n$ are multiplied from the right by scalars in $\mathbb{H}$, and from the left by matrices in $\mathbb{H}^{m \times n}$. With this convention, such matrices define $\mathbb{H}$-linear maps $\mathbb{H}^n \longrightarrow \mathbb{H}^m$. Those who are discomfited by Hamilton's quaternions may ignore them for the time being.

[8] William Rowan Hamilton (1805–1865).

[9] Hermann Günther Graßmann (1809–1877).

Let $F = (f_1, \ldots, f_n)$ be another basis of V. The we write $e_\mu = \sum f_\lambda \alpha^\lambda_\mu$. For $P \in U_E$ and $A := (a^\mu_\nu)$ as above, we then have

$$e_\nu + \sum_{k < \mu \leq n} e_\mu a^\mu_\nu = \sum_{1 \leq \lambda \leq k} f_\lambda \beta^\lambda_\nu + \sum_{k < \lambda \leq n} f_\lambda \gamma^\lambda_\nu$$

with

$$\beta^\lambda_\nu = \alpha^\lambda_\nu + \sum_{k < \mu \leq n} \alpha^\lambda_\mu a^\mu_\nu \quad \text{and} \quad \gamma^\lambda_\nu = \alpha^\lambda_\nu + \sum_{k < \mu \leq n} \alpha^\lambda_\mu a^\mu_\nu.$$

The matrices $B := (\beta^\lambda_\nu) \in \mathbb{K}^{k \times k}$ and $C := (\gamma^\lambda_\nu) \in \mathbb{K}^{(n-k) \times k}$ depend real analytically on A. For $P \in U_E \cap U_F$, B is invertible and

$$\kappa_F(P) = C \cdot B^{-1},$$

so the transition functions are real analytic. Therefore, $\mathcal{A} = ((U_E, \kappa_E))_E$ is a real analytic atlas on $G_k(V)$. For a fixed basis $(e_1, \ldots, e_m)$ of V, the family of charts associated to the bases $(e_{\sigma(1)}, \ldots, e_{\sigma(m)})$ of V, where σ runs over all permutations of $(1, \ldots, m)$, is a finite sub-atlas of $\mathcal{A}$ which satisfies the conditions of Proposition 2.1.13. Therefore $G_k(V)$ together with $\mathcal{A}$ is a real analytic manifold of dimension $dk(n - k)$ with $d = \dim_{\mathbb{R}} \mathbb{K}$ as above. Later we will see that $G_k(V)$ is compact.

6) *Products*: Let M and N be C^k manifolds of dimensions m and n with atlases $\mathcal{A} = ((U_i, x_i))_{i \in I}$ and $\mathcal{B} = ((V_j, y_j))_{j \in J}$. Then the

$$(x_i \times y_j) : U_i \times V_j \longrightarrow U'_i \times V'_j$$

form an $(m + n)$-dimensional C^k atlas $\mathcal{A} \times \mathcal{B}$ on $M \times N$. The topology induced by this atlas on $M \times N$ is the product topology, and is therefore a paracompact Hausdorff topology. So $M \times N$ together with $\mathcal{A} \times \mathcal{B}$ is a C^k manifold of dimension $m + n$.

7) *A strange example*: The set $\mathbb{R}^2$ is covered by the horizontal lines $U_y = \{(x, y) \mid x \in \mathbb{R}\}$, where y runs through all the real numbers. We pair the U_y with the charts

$$\kappa_y : U_y \longrightarrow \mathbb{R}, \quad \kappa_y(x, y) := x.$$

The (U_y, κ_y) then form a one-dimensional real analytic atlas $\mathcal{A}$ on the set $\mathbb{R}^2$, and $\mathbb{R}^2$ together with the real analytic structure induced by $\mathcal{A}$ is a real analytic manifold of dimension 1. To exclude examples like this, many authors, in the definition of a manifold, require the stronger condition that a manifold's topology have a countable basis. However, for the connected components, which in this example are the horizontal lines U_y, both definitions are the same; compare with Exercise 2.7.3.

■

It is a part of the statement of Whitney's Embedding Theorem[10] that the C^k structure of a C^k manifold M contains a real analytic structure if $k \geq 1$ [Wh1, Theorem 1]. For many questions it therefore suffices to consider real analytic manifolds. It is, however, more convenient to restrict ourselves to smooth manifolds; compare, for example, with Lemma 2.1.18. For the sake of simplicity, we will therefore speak of manifolds when we mean smooth manifolds,

$$\text{Convention} : \text{Manifold} = \text{smooth manifold.}$$

In addition to the result of Whitney, it should be mentioned that Kervaire[11] constructed topological manifolds which do not admit a C^1-structure [Ke].

2.1.4 Smooth Maps

Definition 2.1.16

For $k \in \{0, 1, \ldots, \infty\}$ we say that a map $f: M \longrightarrow N$ between two manifolds is C^k if

$$y \circ f \circ x^{-1} : x(U \cap f^{-1}(V)) \longrightarrow V'$$

is C^k for all charts $x: U \longrightarrow U'$ on M and $y: V \longrightarrow V'$ of N. We denote by $C^k(M, N)$ the space of all C^k maps from M to N. For $k = \infty$ we also speak of *smooth maps* and set $\mathcal{F}(M) := C^\infty(M, \mathbb{R})$. See ◘ Fig. 2.3.

Remark 2.1.17

1) Since the composition of C^k maps between open subsets of vector spaces is C^k, to check that a map is C^k, it suffices to find, for every point p in M, charts $x: U \longrightarrow U'$ of M around p and $y: V \longrightarrow V'$ of N around $f(p)$ such that $y \circ f \circ x^{-1}$ is a C^k map.
2) For real analytic manifolds M and N, one can define *real analytic maps* analogous to Definition 2.1.16.

In Exercise 2.7.4 we will discuss examples of smooth maps.

The following lemmas will be important for later constructions. We will not apply them here, but rather wish to use them as examples emphasizing the flexibility of smooth maps. Analogous statements do not hold in the context of real analytic maps: a real analytic map that is constant in a neighborhood of a point is constant on the connected component of that point.

[10] Hassler Whitney (1907–1989).
[11] Michel André Kervaire (1927–2007).

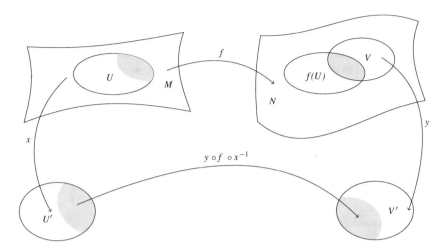

□ Fig. 2.3 A smooth map

Lemma 2.1.18 *For every open neighborhood V about a point p in a manifold M, there is a bump function, that is, a smooth function $f : M \longrightarrow \mathbb{R}$ with $0 \le f \le 1$ and* supp $f \subseteq V$, *such that $f \equiv 1$ locally about p.*

Proof
Choose a chart $x : U \longrightarrow B_1(0) = \{u \in \mathbb{R}^m \mid \|u\| < 1\}$ of M around p with $U \subseteq V$ and $x(p) = 0$, and a smooth function $\varphi : \mathbb{R} \longrightarrow \mathbb{R}$ with $0 \le \varphi \le 1$, $\varphi(r) = 1$ for $r \le 1/3$ and $\varphi(r) = 0$ for $r \ge 2/3$. Then set $f(q) := \varphi(\|x(q)\|)$ for $q \in U$ and $f(q) := 0$ otherwise. Then f is smooth on U and vanishes on $M \setminus x^{-1}(\{u \in \mathbb{R}^m \mid \|u\| \le 2/3\})$. Therefore f is smooth on M. The other properties of f follow directly from the construction.   □

In the following lemma about partitions of unity, the paracompactness of M is an indispensable requirement.

Lemma 2.1.19 (Partition of Unity) *Let (U_i) be an open cover of a manifold M. Then there is a locally finite cover of M by open and relatively compact sets V_j refining (U_i) together with*
1. *charts $x_j : V_j \longrightarrow B_2(0) = \{u \in \mathbb{R}^m \mid \|u\| < 2\}$ and*
2. *smooth functions $\varphi_j : M \longrightarrow \mathbb{R}$ with $0 \le \varphi_j \le 1$,*
such that supp $\varphi_j \subseteq V_j$ *for all j and $\sum_j \varphi_j \equiv 1$.*

The sum over the φ_j is well-defined since every point of M is contained in only finitely many of the V_j, and therefore is contained in only finitely many of the supp φ_j.

Proof
Without loss of generality we may assume that M is connected. Then, by Proposition 2.1.12, M has an exhaustion by compact sets (K_n). Set $K_{-1} = K_0 := \emptyset$ and recursively choose finitely many $V_{nj} \subseteq \mathring{K}_{n+1} \setminus K_{n-2}$ for each $n \ge 1$ with charts $x_{nj} : V_{nj} \longrightarrow B_2(0)$, such

that, for each pair (n, j), there exists an $i \in I$ with $V_{nj} \subseteq U_i$ and such that the compact set $K_n \setminus \overset{\circ}{K}_{n-1}$ is covered by the $x_{nj}^{-1}(B_1(0))$. From this we obtain a locally finite cover of M by open and relatively compact sets V_{nj}, that refines (U_i) and satisfies property 1. We now only need the smooth functions φ_{nj}. To find these, we remember the preceding proof and choose a non-negative smooth function $\psi : \mathbb{R} \longrightarrow \mathbb{R}$ with $\psi > 0$ on $(-1, 1)$ and supp $\psi \subseteq (-2, 2)$. We then set $\psi_{nj}(p) = \psi(\|x_{nj}(p)\|)$ for $p \in V_{nj}$ and $\psi_{nj}(p) = 0$ otherwise. Then the ψ_{nj} are smooth and $\Psi := \sum_{n,j} \psi_{nj}$ is well-defined and smooth, since the covering by the V_{nj} is locally finite. Moreover, $\Psi > 0$, since the $x_{nj}^{-1}(B_1(0))$ cover M and $\psi_{nj} > 0$ on $x_{nj}^{-1}(B_1(0))$. Therefore, the $\varphi_{nj} := \psi_{nj}/\Psi$ satisfy the desired properties. □

In the following we will, in the main, restrict ourselves to smooth maps. On the one hand, real analytic maps are too rigid for our purposes, as indicated above. On the other hand, derivatives of smooth maps are smooth, so we will not have to struggle with the counting of derivatives.

2.1.5 Diffeomorphisms

Definition 2.1.20

A map $f : M \longrightarrow N$ between manifolds is called a *diffeomorphism* if f is a bijection and f and f^{-1} are both smooth. If there is a diffeomorphism $M \longrightarrow N$, M and N are said to be *diffeomorphic*.

Examples of diffeomorphisms will be discussed in Exercise 2.7.5.

Clearly, diffeomorphic manifolds have the same dimension. Connected manifolds of dimension one are diffeomorphic to the line $\mathbb{R}$ or the circle S^1, see [Mi3, pp. 55 ff.]. The diffeomorphism classes of compact connected surfaces are likewise classified, compare with the website of Andrew Ranicki. A central problem in topology is the description of classes of manifolds up to diffeomorphism.

2.1.6 Lie Groups

Definition 2.1.21

A manifold G together with a group structure on G is called a *Lie group*,[12] if the maps

$$\mu : G \times G \longrightarrow G, \quad \mu(g, h) := gh,$$

$$\iota : G \longrightarrow G, \quad \iota(g) := g^{-1},$$

are smooth, where $G \times G$ is seen as carrying the smooth structure described in Example 2.1.15.6.

---
[12]Marius Sophus Lie (1842–1899).

Example 2.1.22

1) A finite-dimensional vector space V with the smooth structure of Example 2.1.15.1 and the operation of addition is a Lie group.

2) For $\mathbb{K} \in \{\mathbb{R}, \mathbb{C}, \mathbb{H}\}$, the *general linear group* $G = \mathrm{Gl}(n, \mathbb{K})$, which consists of all invertible matrices in $\mathbb{K}^{n \times n}$, is an open subset of $\mathbb{K}^{n \times n}$ and is therefore a manifold, compare with Example 2.1.15.2. The multiplication of matrices $G \times G \longrightarrow G$ is clearly smooth. The inverse maps in $\mathrm{Gl}(n, \mathbb{R})$ and $\mathrm{Gl}(n, \mathbb{C})$ are likewise smooth, as one can see from the explicit formula for the inverses of matrices with entries in fields. In the case $\mathbb{K} = \mathbb{H}$, if one views the invertible $(n \times n)$-matrices with entries in $\mathbb{H}$ as $(4n \times 4n)$-matrices with entries in $\mathbb{R}$, and notes that the inverse matrices over $\mathbb{R}$ are also linear over $\mathbb{H}$, it becomes clear that the inverse map for $\mathbb{K} = \mathbb{H}$ is smooth. It follows that $\mathrm{Gl}(n, \mathbb{K})$ is a Lie group.

3) The *Heisenberg group*[13] is the set of all upper triangular matrices in $\mathbb{R}^{3 \times 3}$ of the form

$$\begin{pmatrix} 1 & x & z \\ 0 & 1 & y \\ 0 & 0 & 1 \end{pmatrix}.$$

It is a subgroup of the special linear group $\mathrm{SL}(3, \mathbb{R})$. If we identify the Heisenberg group with $\{(x, y, z) \mid x, y, z \in \mathbb{R}\} = \mathbb{R}^3$ (a global chart), then the multiplication can be written as

$$(x_1, y_1, z_1) \cdot (x_2, y_2, z_2) = (x_1 + x_2, y_1 + y_2, z_1 + z_2 + x_1 y_2).$$

It follows that the Heisenberg group is a Lie group.

∎

Definition 2.1.23

Let G be a Lie group. For $g \in G$, the *left translation* L_g and *right translation* R_g are the maps $G \longrightarrow G$, $L_g(h) := gh$ and $R_g(h) = hg$.

In Exercise 2.7.7, left and right translations will be discussed as diffeomorphisms of Lie groups.

2.2   Tangent Vectors and Derivatives

In analysis, differentiability is defined by the existence of the derivative. Here, we have tacitly assumed this background and spoken about differentiable maps without referring to their derivatives. In the case of differentiable maps in analysis, the derivatives are

---

[13] Werner Karl Heisenberg (1901–1976).

optimal linear approximations and give us information about the local properties of the maps.

How can one define derivatives in our context? Let M and N be manifolds of dimensions m and n respectively and $f : M \longrightarrow N$ be a smooth map. For charts (U, x) of M and (V, y) of N,

$$y \circ f \circ x^{-1} : x(U \cap f^{-1}(V)) \longrightarrow V'$$

is then smooth. Now $x(U \cap f^{-1}(V)) \subseteq \mathbb{R}^m$ and $V' \subseteq \mathbb{R}^n$ are open subsets. Therefore, we can make use of the standard differential $D(y \circ f \circ x^{-1})$ and the standard partial derivatives $\partial_i(y \circ f \circ x^{-1})$ of this map in our discussion of the local properties of f. Then, however, we must retain information about the given chart, since the differential clearly depends on the choice of chart. In practice, this is often what we do. However, it is useful and important to define differentials independently of the choice of charts. To this end, we must first construct vector spaces on which differentials act.

In the following, we denote by M and N manifolds of dimensions m and n respectively. A *curve through* $p \in M$ is a curve $c : I \longrightarrow M$, where I is an open interval with $0 \in I$ and $c(0) = p$.

Proposition and Definition 2.2.1 *Let M be a manifold and p a point in M. We then call two smooth curves c_0 and c_1 though p equivalent, when, with respect to a chart x about p,*

$$\frac{d(x \circ c_0)}{dt}(0) = \frac{d(x \circ c_1)}{dt}(0).$$

This does not depend on the choice of the chart x and defines an equivalence relation on the set of smooth curves through p. We call an equivalence class a tangent vector to M at p (or with basepoint p). We call the set $T_p M$ of all tangent vectors to M at p the tangent space of M at p, and the set $T M = \cup_{p \in M} T_p M$ of all tangent vectors to M the tangent bundle of M.

For $p \in M$ and a smooth curve c through p we denote the equivalence class of c by $[c]$, see ◻ Fig. 2.4. For a chart x around p we then also say that $v := \dot{\sigma}(0) \in \mathbb{R}^m$ represents the tangent vector $[c]$, where $\sigma := x \circ c$. If we do not wish to forget the basepoint p of $[c]$, we instead take as a representative the pair $(x(p), v) \in U' \times \mathbb{R}^m$; compare with Example 2.2.3.1.

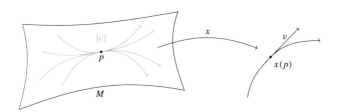

◻ **Fig. 2.4** The equivalence class $[c]$ and the representative v

Proof of Proposition 2.2.1

Let x be a chart on M around p, and let c_0 and c_1 be curves through p, which are equivalent with respect to the chart x around p. Let y be another chart around p and set $\sigma_j := x \circ c_j$ and $\tau_j := y \circ c_j$. Since $\tau_j = (y \circ x^{-1}) \circ \sigma_j$ it follows from the (usual) chain rule that

$$\dot\tau_0(0) = D(y \circ x^{-1})|_{x(p)}(\dot\sigma_0(0))$$
$$= D(y \circ x^{-1})|_{x(p)}(\dot\sigma_1(0)) = \dot\tau_1(0). \qquad \square$$

Remark 2.2.2 Let c be a smooth curve through $p \in M$ and x a chart on M around p. Set $u = x(p)$ and $v = \dot\sigma(0)$ with $\sigma := x \circ c$. Then

$$t \mapsto x^{-1}(u + tv), \quad -\varepsilon < t < \varepsilon,$$

is a curve through p equivalent to c.

Example 2.2.3

1) For an open subset $U \subseteq \mathbb{R}^m$, we choose (U, id) as a chart. Then $T_p U \cong \mathbb{R}^m$ via $T_p U \ni [c] \mapsto \dot c(0) \in \mathbb{R}^m$. If we wish to retain the additional datum of the basepoint p, we set $T_p U \cong \{p\} \times \mathbb{R}^m$; and thereby obtain $TU \cong U \times \mathbb{R}^m$. Accordingly, we can identify the tangent space at a point p and the tangent bundle of an open subset U of a finite dimensional real vector space V, namely $T_p U \cong V$ and $TU \cong U \times V$ respectively.

2) A curve $c \colon I \longrightarrow S^m$ is smooth if and only if it is smooth as a curve in $\mathbb{R}^{m+1}$, that is, as a map $c \colon I \longrightarrow \mathbb{R}^{m+1}$. For a smooth curve $c \colon I \longrightarrow S^m$ through $p \in S^m$ we thereby obtain the usual tangent vector $\dot c(0) \in \mathbb{R}^{m+1}$. From this, we then obtain an identification

$$T_p S^m \cong \{v \in \mathbb{R}^{m+1} \mid \langle p, v \rangle = 0\}.$$

Analogously, we identify the tangent bundle of S^m with

$$T S^m \cong \{(p, v) \in S^m \times \mathbb{R}^{m+1} \mid \langle p, v \rangle = 0\}.$$

3) Let V be a n-dimensional vector space over $\mathbb{K} \in \{\mathbb{R}, \mathbb{C}, \mathbb{H}\}$ and $G_k(V)$ the Grassmannian of k-dimensional, $\mathbb{K}$-linear subspaces of V, compare with Example 2.1.15.5. Let $P \in G_k(V)$ and W be a linear complement of P in V, that is $V = P \oplus W$. Let U be the open set of $Q \in G_k(V)$, which are the graph of a linear map $A_Q \colon P \longrightarrow W$ (which is then uniquely defined by Q). By the definition of the charts on $G_k(V)$, the assignment $\varphi \colon U \longrightarrow \mathrm{Hom}_{\mathbb{K}}(P, W)$, $\varphi(Q) = A_Q$, is then a diffeomorphism. We thereby obtain an identification

$$T_P G_k(V) \cong \mathrm{Hom}_{\mathbb{K}}(P, W), \quad [c] \leftrightarrow \frac{d(\varphi \circ c)}{dt}(0),$$

which is canonical up to the choice of W. If V carries the additional structure of an inner product, we can choose the orthogonal complement $P^\perp$ of P as our complement of P, and obtain $T_P G_k(V) \cong \operatorname{Hom}_\mathbb{K}(P, P^\perp)$.

∎

Remark 2.2.4 We will use the identifications of the tangent spaces in Examples 2.2.3.1–2.2.3.3 throughout.

It is convenient, to expand our notation: For a smooth curve $c\colon I \longrightarrow M$ and $s \in I$, $t \mapsto c(s + t)$, $-\varepsilon < t < \varepsilon$, is a curve through $p := c(s)$. We now set

$$\dot{c}(s) := [t \mapsto c(s + t)] \in T_p M. \tag{2.1}$$

At first this notation seems to conflict with the fact that, for a smooth curve $c\colon I \longrightarrow \mathbb{R}^m$, $\dot{c}(s)$ denotes the usual tangent vector to c in $\mathbb{R}^m$, but on the other hand also denotes the tangent vector in $T_{c(s)}\mathbb{R}^m$. The identification $T_p\mathbb{R}^m \cong \mathbb{R}^m$ as in Example 2.2.3.1 resolves this conflict. In the general case of a smooth curve $c\colon I \longrightarrow M$ and a chart x around $p = c(s)$, $\dot{c}(s)$ is accordingly represented by the vector $\dot{\sigma}(s) \in \mathbb{R}^m$, where $\sigma := x \circ c$.

Proposition and Definition 2.2.5 *Let M and N be manifolds and $f\colon M \longrightarrow N$ be a smooth map. Let p be a point in M, and let c_0 and c_1 be equivalent smooth curves through p. Then $f \circ c_0$ and $f \circ c_1$ are equivalent smooth curves through $f(p)$. We thereby obtain an assignment*

$$f_{*p}\colon T_p M \longrightarrow T_{f(p)} N, \quad [c] \mapsto [f \circ c],$$

which we call the differential *of f at p. We call the induced map $f_*\colon TM \longrightarrow TN$ the* differential *or* pushforward *of f.*

Proof
Let c_0 and c_1 be curves through p with $[c_0] = [c_1]$. Let x and y be charts of M around p and N around $f(p)$ respectively. Let $\sigma_j := x \circ c_j$ and $\tau_j := y \circ f \circ c_j$. Since $\tau_j = (y \circ f \circ x^{-1}) \circ \sigma_j$, it follows from the usual chain rule that

$$\dot{\tau}_0(0) = D(y \circ f \circ x^{-1})|_{x(p)}(\dot{\sigma}_0(0))$$

$$= D(y \circ f \circ x^{-1})|_{x(p)}(\dot{\sigma}_1(0)) = \dot{\tau}_1(0). \qquad \square$$

We write f_{*p}, when we consider the differential of f at a point p and wish to retain the datum of the point p in our notation. In many cases, we will stick with the simpler f_*, as f_{*p} is merely the restriction of f_* to $T_p M$. The relationship between f_{*p} and $D(y \circ f \circ x^{-1})|_{x(p)}$ is clarified in the left-hand diagram in (2.4) below.

Chain Rule 2.2.6 *For smooth maps* $f\colon M \longrightarrow N$ *and* $g\colon N \longrightarrow P$ *between manifolds,* $(g \circ f)_* = g_* \circ f_*$.

Proof

Let $p \in M$ and c be a smooth curve through p. Then $f \circ c$ is a smooth curve through $f(p)$ and

$$(g \circ f)_{*p}([c]) = [(g \circ f) \circ c] = [g \circ (f \circ c)]$$

$$= g_{*f(p)}([f \circ c]) = g_{*f(p)}(f_{*p}([c])).$$

Therefore $(g \circ f)_{*p} = g_{*f(p)} \circ f_{*p}$ for all $p \in M$, as claimed.   □

We are still missing a linear structure on the tangent spaces such that the differentials become linear maps. To this end, let $p \in M$ and let x be a chart on M around p. By Definition 2.2.1 and Remark 2.2.2, the map

$$dx(p)\colon T_p M \longrightarrow \mathbb{R}^m, \quad dx(p)([c]) := \frac{d(x \circ c)}{dt}(0), \tag{2.2}$$

is a bijection. We now describe a vector space structure on $T_p M$ such that $dx(p)$ becomes an isomorphism:

Proposition and Definition 2.2.7 *Let* $[c_0], [c_1], [c_2] \in T_p M$ *and* $\alpha \in \mathbb{R}$. *Set* $\sigma_j := x \circ c_j$. *Then the rules*

$$[c_0] + [c_1] := [c_2] \iff \dot{\sigma}_0(0) + \dot{\sigma}_1(0) = \dot{\sigma}_2(0)$$
$$\alpha \cdot [c_0] := [c_1] \iff \alpha \cdot \dot{\sigma}_0(0) = \dot{\sigma}_1(0) \tag{2.3}$$

are independent of the chart x. *With respect to this structure,* $T_p M$ *is a real vector space such that* $dx(p)\colon T_p M \longrightarrow \mathbb{R}^m$ *becomes an isomorphism. The differential* $f_{*p}\colon T_p M \longrightarrow T_q N$, $q := f(p)$, *of a smooth map* $f\colon M \longrightarrow N$ *is linear with respect to this structure.*

Proof

We assume that the right-hand sides in (2.3) hold. Let y be another chart on M around p, and set $\tau_j := y \circ \sigma_j$. Then $\tau_j = (y \circ x^{-1}) \circ \sigma_j$. From the usual chain rule, we obtain

$$\dot{\tau}_0(0) + \dot{\tau}_1(0) = D(y \circ x^{-1})|_{x(p)}(\dot{\sigma}_0(0)) + D(y \circ x^{-1})|_{x(p)}(\dot{\sigma}_1(0))$$

$$= D(y \circ x^{-1})|_{x(p)}(\dot{\sigma}_0(0) + \dot{\sigma}_1(0))$$

$$= D(y \circ x^{-1})|_{x(p)}(\dot{\sigma}_2(0)) = \dot{\tau}_2(0),$$

where we have used that the usual differential $D(y \circ x^{-1})|_{x(p)}$ is additive. From this it follows that the addition on $T_p M$ is independent of the choice of chart. One proves analogously that multiplication by real scalars is well-defined.

Now let $f : M \longrightarrow N$ be a smooth map and y a chart on N around $f(p)$. Then for $\sigma_j = x \circ c_j$ as above, and $\tau_j := y \circ f \circ c_j$, we have

$$\dot{\tau}_0(0) + \dot{\tau}_1(0) = D(y \circ f \circ x^{-1})|_{x(p)}(\dot{\sigma}_0(0)) + D(y \circ f \circ x^{-1})|_{x(p)}(\dot{\sigma}_1(0))$$

$$= D(y \circ f \circ x^{-1})|_{x(p)}(\dot{\sigma}_0(0) + \dot{\sigma}_1(0))$$

$$= D(y \circ f \circ x^{-1})|_{x(p)}(\dot{\sigma}_2(0)) = \dot{\tau}_2(0),$$

where we are now using that $D(y \circ f \circ x^{-1})|_{x(p)}$ is additive. From this it follows that f_{*p} is additive. One proves homogeneity with respect to multiplication by real scalars analogously.

$\square$

Consistent with (2.2), we expand the catalog of derivatives with one further variant, namely, the *differential*[14] $df : TM \longrightarrow V$ of a smooth map $f : M \longrightarrow V$, where V is a finite-dimensional real vector space. We identify $T_{f(p)}V \cong V$ as in Example 2.2.3.1 and describe $df(p)$, in that we require that the right hand diagram commutes:

$$(2.4)$$

By definition, then, $df(p)$ is a linear map and

$$df(p)([c]) = \frac{d(f \circ c)}{dt}(0) \in V, \quad [c] \in T_p M. \tag{2.5}$$

Instead of $df(p)$, we also sometimes write $df|_p$. When there is no danger of misunderstanding, we write df instead of $df(p)$ for simplicity.

Lemma 2.2.8 *Let M be a manifold, $p \in M$, and $x : U \longrightarrow U'$ be a chart on M around p. Then there is an open neighborhood V around p in U, such that, for every $f \in \mathcal{F}(M)$, there exist smooth functions $f_i : V \longrightarrow \mathbb{R}$, $1 \leq i \leq m$, with*[15]

$$f|_V = f(p) + (x^i - x^i(p))f_i,$$

where $x = (x^1, \ldots, x^m)$ and $f_i(p) = \partial_i(f \circ x^{-1})(x(p))$.

---

[14] Our terminology is not uncommon, but there are multiple conventions in the literature.

[15] Recall the Einstein summation convention.

Proof

Let $f \in \mathcal{F}(M)$, $u_0 := x(p)$, and let $V' \subseteq U'$ be an open ball around u_0. Then for $u \in V'$ and $\varphi := f \circ x^{-1}$, we have

$$\varphi(u) - \varphi(u_0) = \int_0^1 \frac{d}{dt} \varphi(tu + (1-t)u_0) dt$$

$$= (u^i - u_0^i) \int_0^1 (\partial_i \varphi)(tu + (1-t)u_0) dt.$$

We now define smooth functions $\varphi_i : V' \longrightarrow \mathbb{R}$ via

$$\varphi_i(u) := \int_0^1 (\partial_i \varphi)(tu + (1-t)u_0) dt.$$

Then the $f_i := \varphi_i \circ x$ are smooth on $V = x^{-1}(V') \subseteq U$ and provide the desired representation of f.   □

Reversing our perspective on the definition of the differential, we now view tangent vectors as directional derivatives.

Proposition and Definition 2.2.9 *Let M be a manifold, $p \in M$, and $v \in T_p M$. Let c be a smooth curve through p with $[c] = v$. Then the derivative in the v-direction, which we also denote by v:*

$$v : \mathcal{F}(M) \longrightarrow \mathbb{R}, \quad v(f) = \frac{d(f \circ c)}{dt}(0), \tag{2.6}$$

is well-defined, $\mathbb{R}$-linear and satisfies the product rule

$$v(f \cdot g) = v(f) \cdot g(p) + f(p) \cdot v(g). \tag{2.7}$$

Conversely, for every $\mathbb{R}$-linear map $a : \mathcal{F}(M) \longrightarrow \mathbb{R}$ that satisfies the product rule (2.7), there is a $v \in T_p M$, such that a is the derivative in the v-direction as in (2.6). The canonical $\mathbb{R}$-vector space structure on the space of such maps corresponds under this association to the vector space structure on $T_p M$ defined in Proposition 2.2.7. In the sense of (2.6), the differential of a smooth map $h : M \longrightarrow N$ evaluates to

$$h_*(v)(f) = v(f \circ h), \quad v \in TM \text{ and } f \in \mathcal{F}(N). \tag{2.8}$$

Proof

We only prove the characterization of the linear maps a that satisfy the product rule. The proof of the remaining assertions of the theorem is left as an exercise to the reader. Let $a : \mathcal{F}(M) \longrightarrow \mathbb{R}$ be a $\mathbb{R}$-linear map, that satisfies the product rule (2.7). We begin with two preliminaries.

First preliminary: Let $f_1, f_2 \in \mathcal{F}(M)$ be functions, that agree on an (open) neighborhood W of p in M. Let $\varphi \in \mathcal{F}(M)$ be a bump function with respect to p and W as in Lemma 2.1.18. Then $\varphi f_1 = \varphi f_2$ and therefore

$$a(\varphi) f_1(p) + a(f_1) = a(\varphi f_1) = a(\varphi f_2) = a(\varphi) f_2(p) + a(f_2).$$

Now $f_1(p) = f_2(p)$, so $a(f_1) = a(f_2)$.

Second preliminary: For the constant function 1, we obtain

$$a(1) = a(1 \cdot 1) = a(1) \cdot 1 + 1 \cdot a(1) = 2 \cdot a(1),$$

so $a(1) = 0$. From the linearity of a it therefore also follows that $a(f) = 0$ for all constant functions f on M.

Now let $f: M \longrightarrow \mathbb{R}$ be smooth. For a chart (U, x) on M about p and a bump function φ as in Lemma 2.1.18 with $\operatorname{supp} \varphi \subseteq U$, f and $\varphi^2 \cdot f$ are smooth functions, that agree on a neighborhood of p in M. From this and the first preliminary above, it follows that $a(f) = a(\varphi^2 \cdot f)$. Further, by Lemma 2.2.8

$$\varphi^2 \cdot f = \varphi^2 \cdot f(p) + (\varphi \cdot (x^i - x^i(p)))(\varphi \cdot f_i)$$

with $f_i(p) = \partial_i (f \circ x^{-1})(x(p))$. Now $\operatorname{supp} \varphi \subseteq U$, and thus the functions $\varphi \cdot (x^i - x^i(p))$ and $\varphi \cdot f_i$ are smooth on the whole of M when they are extended by 0 on $M \setminus U$. Therefore, the functions thus continued are in the domain of a, and

$$a(\varphi^2 \cdot f) = a(\varphi \cdot (x^i - x^i(p))) \cdot f_i(p)$$
$$= a^i \partial_i (f \circ x^{-1})(x(p))$$

with $a^i := a(\varphi \cdot (x^i - x^i(p)))$. Therefore

$$a(f) = \frac{d(f \circ c)}{dt}(0) \quad \text{with} \quad c = c(t) = x^{-1}(x(p) + t a^i e_i). \qquad \square$$

Remark 2.2.10 Clearly, the derivative $v(f)$ in the v-direction as in (2.6) is also well-defined for maps $f \in C^1(U, V)$, where U is an open neighborhood of p in M and V is a finite dimensional vector space over $\mathbb{K} \in \{\mathbb{R}, \mathbb{C}, \mathbb{H}\}$.

We now return to the computation of differentials with respect to local coordinates. To this end, let $x: U \longrightarrow U'$ be a chart on M. For $p \in U$ we obtain the tangent vector corresponding to the coordinate direction

$$\frac{\partial}{\partial x^i}(p) = \frac{\partial}{\partial x^i}\bigg|_p := [t \mapsto x^{-1}(x(p) + t e_i)] \in T_p M, \qquad (2.9)$$

where $e_i \in \mathbb{R}^m$ denotes the i-th unit vector, $1 \le i \le m$; compare also with Remark 2.2.2. In the sense of (2.6) or Remark 2.2.10, therefore,

$$\frac{\partial f}{\partial x^i}(p) := \frac{\partial}{\partial x^i}\Big|_p (f) = \partial_i(f \circ x^{-1})\Big|_{x(p)} \tag{2.10}$$

is the i-th partial derivative of $f \circ x^{-1}$ at the point $x(p)$. For the components x^j of x, which are smooth functions on U, we have

$$\frac{\partial x^j}{\partial x^i}(p) = \delta_i^j, \quad 1 \le i, j \le m. \tag{2.11}$$

From Proposition 2.2.7 the differential $\mathbb{R}^m \ni v \mapsto [x^{-1}(x(p) + tv)] \in T_p M$ is an isomorphism, so the

$$\frac{\partial}{\partial x^1}\Big|_p, \dots, \frac{\partial}{\partial x^m}\Big|_p \tag{2.12}$$

form a basis of $T_p M$, the *standard basis* of $T_p M$ with respect to the chart x.

Proposition 2.2.11 *For $v \in T_p M$*[16]

$$v = v(x^i)\frac{\partial}{\partial x^i}\Big|_p.$$

Proof

For $v \in T_p M$ and $f \in \mathcal{F}(M)$ we compute with a representation of f as in Lemma 2.2.8:

$$v(f) = v(x^i) f_i(p) = v(x^i)\frac{\partial f}{\partial x^i}(p). \qquad \square$$

Corollary 2.2.12 (Transformation Rules)

1. *For charts x and y on M about p,*

$$\frac{\partial}{\partial y^i}\Big|_p = \frac{\partial x^j}{\partial y^i}(p)\frac{\partial}{\partial x^j}\Big|_p.$$

2. *Let $f: M \longrightarrow N$ be smooth, and let x and y be charts on M about p and N about $f(p)$. With $f^j := y^j \circ f$, then,*

$$f_{*p}\Big(\frac{\partial}{\partial x^i}\Big|_p\Big) = \frac{\partial f^j}{\partial x^i}(p)\frac{\partial}{\partial y^j}\Big|_{f(p)}.$$

---

[16]In the Einstein summation convention, the index i in $\partial/\partial x^i$ is counted as a lower index.

If one does not write out the points p and $f(p)$ in the formulas, they become more legible:

$$\frac{\partial}{\partial y^i} = \frac{\partial x^j}{\partial y^i}\frac{\partial}{\partial x^j} \quad \text{and} \quad f_*\left(\frac{\partial}{\partial x^i}\right) = \frac{\partial f^j}{\partial x^i}\frac{\partial}{\partial y^j}. \tag{2.13}$$

These formulas are to be understood with the corresponding points substituted in.

Inverse Function Theorem 2.2.13 *If $f_{*p}\colon T_pM \longrightarrow T_{f(p)}N$ is an isomorphism, then there are open neighborhoods U of p in M and V of $f(p)$ in N, such that $f\colon U \longrightarrow V$ is a diffeomorphism.*

Proof

Let $x\colon U \longrightarrow U'$ be a chart on M around p and $y\colon V \longrightarrow V'$ a chart on N around $f(p)$. By refining U, we can assume $f(U) \subseteq V$. Then $\varphi := (y \circ f \circ x^{-1})\colon U' \longrightarrow V'$ is a smooth map such that $D\varphi|_{x(p)}$ is invertible. Therefore, there are neighborhoods $\hat{U}'$ of $x(p)$ in U' and $\hat{V}'$ of $y(f(p)) = \varphi(x(p))$ in V', such that $\varphi\colon \hat{U}' \longrightarrow \hat{V}'$ is a diffeomorphism. With $\hat{U} = x^{-1}(\hat{U}')$ and $\hat{V} = y^{-1}(\hat{V}')$, $f = (y^{-1} \circ \varphi \circ x)\colon \hat{U} \longrightarrow \hat{V}$ is then a diffeomorphism. □

We call $p \in M$ a *regular point* of a smooth map $f\colon M \longrightarrow N$, if f_{*p} is surjective, and $q \in N$ a *regular value* (of f), if all $p \in f^{-1}(q)$ are regular points. We say that f is a *submersion*, if all $p \in M$, or, equivalently, all $q \in N$ are regular. Warning: points in $N \setminus \text{im } f$ are regular.

We say that f is an *immersion* if all f_{*p}, $p \in M$, are injective. If, additionally, $f\colon M \longrightarrow f(M)$ is a homeomorphism, where $f(M)$ is seen with the relative topology, then we call f an *embedding*.

Occasionally, one refers to immersions as regular maps—in particular, a curve $c\colon I \longrightarrow M$ is called *regular*, if $\dot{c}(t) \neq 0$ for all $t \in I$.

The *rank* of f at $p \in M$ is the dimension of the image of f_{*p}. If x is a chart on M about p and y is a chart on N about $f(p)$, then the rank of f at p equals the rank $y \circ f \circ x^{-1}$ at $x(p)$, since x and y are diffeomorphisms.

2.3 Submanifolds

Definition 2.3.1

A subset $L \subseteq M$ is called a *submanifold of M of dimension ℓ* (or *codimension $m-\ell$*), if, for every $p \in L$, there is an *adapted chart* $x\colon U \longrightarrow U' \times U''$ on M around p, where $U' \subset \mathbb{R}^\ell$ and $U'' \subset \mathbb{R}^{m-\ell}$ are open with $0 \in U''$ and $x(U \cap L) = U' \times \{0\}$; see ◼ Fig. 2.5. Submanifolds of codimension 1 are also called *hypersurfaces*.

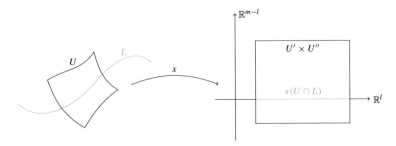

■ **Fig. 2.5** An adapted chart

Proposition 2.3.2 *Submanifolds* $L \subseteq M$ *are manifolds:*
1. *The restriction* $x \colon U \cap L \longrightarrow U' \times \{0\} \cong U'$ *of the adapted charts* $x \colon U \longrightarrow U' \times U''$ *define a smooth atlas on* L, *whose induced topology agrees with the the relative topology of* L *as a subset of* M.
2. *The inclusion* $i \colon L \longrightarrow M$ *is an embedding. In particular, for all* $p \in L$ *and adapted charts* x *around* p, *we have*

$$T_p L \cong \operatorname{im} i_{*p} = \cap_{j>\ell} \ker dx^j(p) \subseteq T_p M.$$

3. *A map* $f \colon L \longrightarrow N$ *is smooth if and only if, for every* $p \in L$, *there is an open neighborhood* U *of* p *in* M *and a smooth map* $g \colon U \longrightarrow N$ *such that* $f|_{U \cap L} = g|_{U \cap L}$. *Then* $f_{*p} = g_{*p}|_{T_p L}$ *with respect to the identification* $T_p L \cong \operatorname{im} i_{*p}$.

Remark 2.3.3 By the Whitney Embedding Theorem [Wh1, Theorem 1], every m-dimensional manifold is diffeomorphic to a (real analytic) submanifold of $\mathbb{R}^{2m+1}$.

Proof of Proposition 2.3.2
We show only that the transition functions are smooth. We leave the proof of the remaining claims as an exercise.

Let $x \colon U \longrightarrow U' \times U''$ and $y \colon V \longrightarrow V' \times V''$ be adapted charts for L. Since x and y are charts on M, $x \circ y^{-1}$ is smooth. Moreover,

$$x(U \cap V \cap L) = x(U \cap V) \cap (U' \times \{0\})$$

and $y(U \cap V \cap L)$ are open in $U' \times \{0\} \cong U'$ and $V' \times \{0\} \cong V'$ respectively. Therefore the restriction

$$x \circ y^{-1} \colon x(U \cap V \cap L) \longrightarrow y(U \cap V \cap L)$$

is smooth as a map between open subsets of $\mathbb{R}^\ell$, $\ell = \dim L$.   □

Corollary 2.3.4 *Let* $L \subseteq M$ *be a submanifold and* $g \colon M \longrightarrow N$ *be smooth. Then* $f := g|_L \colon L \longrightarrow N$ *is smooth, and* $f_{*p} = g_{*p}|_{T_p L}$ *for all* $p \in L$.   □

Submanifolds arise as level sets of smooth maps: a level set of a smooth map is a submanifold provided that the differential of the map is surjective at every point of the level set. For the following discussion, we recall that the rank of a smooth function $f: M \longrightarrow N$ at $p \in M$ is defined to be the rank of the differential f_{*p}.

Proposition 2.3.5 *Let $f: M \longrightarrow N$ be a smooth map. Then:*
1. *Suppose f has rank r at $p \in M$. Then, for every chart (y, V) around $f(p)$ with $y(f(p)) = 0$, there is a chart (x, U) around p with $x(p) = 0$, such that, after possible renumbering of the components of y, we have*

$$\varphi(u^1, \ldots, u^m) = (u^1, \ldots, u^r, \varphi^{r+1}(u), \ldots, \varphi^n(u))$$

with $\varphi := y \circ f \circ x^{-1}$, $\varphi^j(0) = 0$ and $(\partial_i \varphi^j)(0) = 0$ for $i, j > r$.
2. *If f has rank r on a neighborhood of p, then there are charts (x, U) around p with $x(p) = 0$ and (y, V) around $f(p)$ with $y(f(p)) = 0$, such that*

$$\varphi(u^1, \ldots, u^m) = (u^1, \ldots, u^r, 0, \ldots, 0).$$

Proof

Let $\hat{x}$ be a chart around p with $\hat{x}(p) = 0$ and $\hat{\varphi} := y \circ f \circ \hat{x}^{-1}$. After possible renumbering of the components of $\hat{x}$ and y we can assume that the matrix

$$\left((\partial_i \hat{\varphi}^j)(0) \right)_{1 \le i, j \le r}$$

is invertible. We set $x^j := y^j \circ f$ for $1 \le j \le r$ and $x^j := \hat{x}^j$ for $r < j \le m$. Then $x(p) = 0 \in \mathbb{R}^m$. Furthermore,

$$\left((\partial_i (x^j \circ \hat{x}^{-1}))(0) \right) = \begin{pmatrix} (\partial_i \hat{\varphi}^j)(0) & * \\ 0 & 1 \end{pmatrix},$$

and so x has rank m at p. By the Inverse Function Theorem, x is a local diffeomorphism around p: there is a neighborhood U about p in M and a neighborhood U' of 0 in $\mathbb{R}^m$, such that $x: U \longrightarrow U'$ is a chart on M. By the definition of x,

$$\varphi(u) = (y \circ f \circ x^{-1})(u^1, \ldots, u^m)$$
$$= (u^1, \ldots, u^r, \varphi^{r+1}(u), \ldots, \varphi^n(u)),$$

where $\varphi^{r+1}, \ldots, \varphi^n$ are smooth functions on U' with $\varphi^j(0) = 0$. The Jacobian matrix[17] φ on U' is therefore

$$\begin{pmatrix} 1 & 0 \\ * & (\partial_i \varphi^j)_{i, j > r} \end{pmatrix}.$$

---

[17]Carl Gustav Jacob Jacobi (1804–1851).

Since φ has rank r at $u = 0$, it follows that $(\partial_i \varphi^j)(0) = 0$ for all $i, j > r$. The chart x on M about p therefore satisfies the conditions in (1).

Under the assumptions in (2), it is further required that $\partial_i \varphi^j \equiv 0$, $i, j > r$, on a neighborhood of 0. If we refine U, we can assume that $U' = (-\varepsilon, \varepsilon)^m$, and that the partial derivatives $\partial_i \varphi^j$ vanish on U' for all $i, j > r$. Then it follows that

$$\varphi^j(u) = \varphi^j(u^1, \ldots, u^r), \quad r < j \le m;$$

that is, the φ^j do not depend on $u^{r+1}, \ldots, u^m$. Without loss of generality, we can assume, following (1), that $V' = y(V)$ is of the form $(-\varepsilon, \varepsilon)^r \times (-\delta, \delta)^{n-r}$. We now change the chart y of N and set $\hat{y}^j = y^j$ for $1 \le j \le r$ and $\hat{y}^j = y^j - \varphi^j(y^1, \ldots, y^r)$ otherwise. Then,

$$\left(\frac{\partial \hat{y}^j}{\partial x^i} \right) = \begin{pmatrix} 1 & 0 \\ * & 1 \end{pmatrix}.$$

By the Inverse Function Theorem, $\hat{y}$ is a local diffeomorphism around $f(p)$, and thus is a chart of N on an open neighborhood $\hat{V}$ of $f(p)$. By definition, then

$$(\hat{y} \circ f \circ x^{-1})(u^1, \ldots, u^m) = (u^1, \ldots, u^r, 0, \ldots, 0). \qquad \square$$

Corollary 2.3.6 *If f has rank r on a neighborhood of $L = f^{-1}(q)$, then L is a submanifold of M of codimension r with $T_p L \cong \ker f_{*p}$ for all $p \in L$.*

Proof
For $p \in M$ there are, by Proposition 2.3.5, charts (U, x) on M around p and y on N around $f(p) = q$ with $x(p) = 0$ and $y(q) = 0$, such that

$$(y \circ f \circ x^{-1})(u^1, \ldots, u^m) = (u^1, \ldots, u^r, 0 \ldots, 0).$$

By refining U, we can assume that $x(U) = (-\varepsilon, \varepsilon)^m$. With $U' = (-\varepsilon, \varepsilon)^{m-r}$ and $U'' = (-\varepsilon, \varepsilon)^r$, then, up to exchanging the factors U' and U'', $x : U \longrightarrow U'' \times U'$ is an adapted chart on M around p with $\ker f_{*p} = \cap_{j \le r} \ker dx^j(p)$. $\qquad \square$

Corollary 2.3.7 (Implicit Function Theorem) *If f_{*p} is surjective, then for every chart y on N around $f(p)$ with $y(f(p)) = 0$ there is a chart x on M around p with $x(p) = 0$, such that*

$$(y \circ f \circ x^{-1})(u^1, \ldots, u^m) = (u^1, \ldots, u^n).$$

If, in particular, q is a regular value of f, then $L = f^{-1}(q)$ is a submanifold of M of dimension $m - n$ with $T_p L \cong \ker f_{*p}$ for all $p \in L$.

Proof

The condition on p that f_{*p} be surjective is open: if it is fulfilled at $p \in M$, it is also fulfilled on a neighborhood of p. □

Example 2.3.8

1) Let B be a symmetric bilinear form on a m-dimensional real vector space V, and $Q(x) :=$ $B(x, x)$ the corresponding quadratic form. If $\alpha \in \mathbb{R} \setminus \{0\}$ lies in the image of Q, then α is a regular value of Q, and the *quadric*

$$Q_\alpha = \{x \in V \mid Q(x) = \alpha\}$$

is therefore a hypersurface in V. For all $x \in Q_\alpha$ there is a canonical isomorphism

$$T_x Q_\alpha \cong \ker dQ(x) \cong \{y \in V \mid Q(x, y) = 0\} \subseteq V,$$

where we at the second $\cong$ identify $T_x V$ with V as usual. In the special case of the Euclidean scalar product on $V = \mathbb{R}^m$ and $\alpha = 1$, we obtain the unit sphere. Compare with Examples 2.1.2.2 and 2.2.3.2.

2) For $\mathbb{K} \in \{\mathbb{R}, \mathbb{C}\}$, $\det: \mathrm{Gl}(n, \mathbb{K}) \longrightarrow \mathbb{K}$ is smooth with differential

$$D\det|_A(B) = \frac{d}{dt} \det(A + tB)|_{t=0}$$

$$= \frac{d}{dt}(\det(A)\det(E + tA^{-1}B))|_{t=0}$$

$$= \det(A)\,\mathrm{tr}(A^{-1}B),$$

where E denotes the identity matrix. Therefore det is of constant, maximal rank $\dim_\mathbb{R} \mathbb{K}$ and therefore, the *special linear group* $\mathrm{Sl}(n, \mathbb{K}) = \{A \in \mathrm{Gl}(n, \mathbb{K}) \mid \det A = 1\}$ is a *Lie subgroup* of $\mathrm{Gl}(n, \mathbb{K})$, that is, simultaneously a submanifold and a subgroup, and therefore, in particular, is itself a Lie group.

3) Let $\mathbb{K} \in \{\mathbb{R}, \mathbb{C}, \mathbb{H}\}$ and $G = \{A \in \mathbb{K}^{n \times n} \mid A^*A = E\}$, where A^* denotes the conjugate-transpose matrix of A and E the identity matrix (of the appropriate size). The set of these matrices is familiar from linear algebra, at least in the cases $\mathbb{K} \in \{\mathbb{R}, \mathbb{C}\}$. For $\mathbb{K} = \mathbb{R}$, G is called the *orthogonal group*, for $\mathbb{K} = \mathbb{C}$, the *unitary group* and for $\mathbb{K} = \mathbb{H}$ the *symplectic group*, denoted by $O(n)$, $U(n)$ and $Sp(n)$ respectively. All three consist of invertible matrices and are, under matrix multiplication, compact subgroups of the respective general linear groups. Furthermore,

$$O(1) = \{\pm 1\} = S^0, \quad U(1) = S^1, \quad Sp(1) = S^3.$$

We set

$$H_n(\mathbb{K}) := \{A \in \mathbb{K}^{n \times n} \mid A^* = A\}.$$

Then $H_n(\mathbb{K})$ is a real vector space of dimension $n + dn(n-1)/2$, $d = \dim_{\mathbb{R}} \mathbb{K}$, and G is the level set of the smooth map

$$f : \mathbb{K}^{n \times n} \mapsto H_n(\mathbb{K}), \quad f(A) = A^* A.$$

For $A, B \in \mathbb{K}^{n \times n}$, we have

$$df(A)(B) = \frac{d}{dt}((A + tB)^*(A + tB))|_{t=0} = A^* B + B^* A.$$

For $A \in G$ and $C \in \mathbb{K}^{n \times n}$, therefore,

$$df(A)(AC) = \begin{cases} 0 & \text{if } C^* = -C, \\ 2C & \text{if } C^* = C. \end{cases}$$

In particular, $E \in H_n(\mathbb{K})$ is a regular value of f, therefore G is a submanifold of $\mathbb{K}^{n \times n}$ of dimension $(d-1)n + dn(n-1)/2$. Now, G is a subgroup of $\mathrm{Gl}(n, \mathbb{K})$, and therefore also a Lie subgroup of $\mathrm{Gl}(n, \mathbb{K})$. For $\mathbb{K} = \mathbb{R}$, $\mathbb{K} = \mathbb{C}$, or $\mathbb{K} = \mathbb{H}$, $T_E G$ is also denoted by $\mathfrak{so}(n)$, $\mathfrak{u}(n)$, or $\mathfrak{sp}(n)$ respectively.

The orthogonal group has two connected components; the connected component $\mathrm{SO}(n)$ of the identity consists of the orientation-preserving orthogonal transformations of $\mathbb{R}^n$.

In the case of the unitary group, $\det \colon \mathrm{U}(n) \longrightarrow S^1$ is a smooth homomorphism with maximal rank 1. Therefore, the *special unitary group* $\mathrm{SU}(n) = \{A \in \mathrm{U}(n) \mid \det A = 1\}$ is a submanifold of $\mathrm{U}(n)$ of codimension 1 and thus a Lie subgroup of $\mathrm{U}(n)$ and $\mathrm{Gl}(n, \mathbb{C})$. ∎

Corollary 2.3.9 (of Proposition 2.3.5) *If f_{*p} is injective, then there are charts x around p and y around $f(p)$ with*

$$(y \circ f \circ x^{-1})(u_1, \ldots, u_m) = (u_1, \ldots, u_m, 0, \ldots, 0).$$

In particular, there is an open neighborhood U of p in M, such that $L = f(U)$ is a submanifold of N and $f : U \longrightarrow L$ is a diffeomorphism.

Proof

The condition on p that f_{*p} be injective is open.   □

2.4   Tangent Bundles and Vector Fields

Our next goal is the construction of an atlas on the tangent bundle TM of a manifold M; see Definition 2.2.1. For $A \subseteq M$ we set $TM|_A := \cup_{p \in A} T_p M$.

Proposition 2.4.1 *For a chart $x : U \longrightarrow U' \subseteq \mathbb{R}^m$ on M, the differential*

$$x_* : TM|_U \longrightarrow U' \times \mathbb{R}^m, \quad x_*([c]) := (x(c(0)), dx(c(0))([c])), \tag{2.14}$$

is a bijection, where we identify TU' with $U' \times \mathbb{R}^m$ as is Example 2.2.3.1. If $y : V \longrightarrow V'$ is another chart on M, then the transition functions are given by

$$(y_* \circ x_*^{-1})(u, v) = ((y \circ x^{-1})(u), D(y \circ x^{-1})|_u(v)). \tag{2.15}$$

With respect to the charts $(TM|_U, x_*)$, TM is a manifold, that is, these charts comprise a smooth atlas of TM, whose associated topology is Hausdorff and paracompact. The projection onto the basepoint,

$$\pi : TM \longrightarrow M, \quad \pi(v) := p \quad \text{for } v \in T_p M, \tag{2.16}$$

is a submersion. Furthermore, the differential $f_* : TM \longrightarrow TN$ of a smooth map $f : M \longrightarrow N$ is smooth with respect to these smooth structures.

Proof

By Proposition 2.2.7, $x_* : TM|_U \longrightarrow U' \times \mathbb{R}^m$ is bijective. The claim about transition functions follows from transformation rule 2.2.12.1. We leave the proof of the other claims as an exercise. □

The atlas on TM that we constructed in Proposition 2.4.1 is compatible with the linear structure on the tangent spaces. We will expand on this with the next proposition.

Proposition and Definition 2.4.2 *A vector field on M is a map $X : M \longrightarrow TM$ with $\pi \circ X = \mathrm{id}_M$. For a chart x on M with associated chart x_* on TM as in Proposition 2.4.1, a vector field X on M is of the form*

$$(x_* \circ X \circ x^{-1})(u) = (u, \xi(u)), \quad u \in U', \tag{2.17}$$

over U, and X is smooth on U if and only if the principal part ξ of X with respect to x is smooth. For vector fields X and Y and a real function f on M, $X + Y$ and fX, defined via

$$(X + Y)(p) := X(p) + Y(p) \quad \text{and} \quad (fX)(p) := f(p)X(p), \tag{2.18}$$

are again vector fields on M. If X, Y and f are smooth, then so are $X+Y$ and fX. Therefore, the set $\mathcal{V}(M)$ of smooth vector fields on M becomes a vector space over $\mathbb{R}$ and a module over $\mathcal{F}(M)$. □

For an open subset $W \subseteq M$, a smooth vector field X on W, and $\varphi \in \mathcal{F}(W)$, let $X\varphi = X(\varphi) \in \mathcal{F}(W)$ be defined by

$$(X\varphi)(p) := X_p(\varphi). \tag{2.19}$$

The map $\mathcal{F}(W) \longrightarrow \mathcal{F}(W), \varphi \mapsto X\varphi$, is a *derivation* on the ring $\mathcal{F}(W)$; that is, for all $\varphi, \psi \in \mathcal{F}(W)$

$$X(\varphi \cdot \psi) = X(\varphi) \cdot \psi + \varphi \cdot X(\psi). \tag{2.20}$$

Zeros of smooth (or continuous) vector fields have topological relevance: The Hedgehog Theorem says that every continuous vector field on the sphere S^2 has a zero. In particular, the tangent bundle of S^2 is not trivial in the sense of Example 2.5.2.1. The general Poincaré-Hopf Theorem[18] has the consequence that the so-called Euler characteristic[19] of a compact manifold vanishes if it possesses a vector field without zeros.

Example 2.4.3
On the sphere $S^{2n-1} \subseteq \mathbb{C}^n$, $x \mapsto ix$ is a smooth vector field (with respect to the usual identification of TS^{2n-1}) without zeros.   ∎

2.4.1 Lie Bracket

For $X, Y \in \mathcal{V}(M)$ we define a new smooth vector field on M, the *Lie bracket* $[X, Y]$ of X and Y, via

$$[X, Y]_p(\varphi) := X_p(Y\varphi) - Y_p(X\varphi). \tag{2.21}$$

The Lie bracket corresponds to the commutator of the aforementioned derivations. Here, we identify tangent vectors with directional derivatives in the sense of Proposition 2.2.9. For this, we must show that $[X, Y]_p$ satisfies the product rule (2.7):

$$\begin{aligned}
[X, Y]_p(\varphi\psi) &= X_p(Y(\varphi\psi)) - Y_p(X(\varphi\psi)) \\
&= X_p((Y\varphi)\psi + \varphi(Y\psi)) - Y_p((X\varphi)\psi + \varphi(X\psi)) \\
&= (X_p(Y\varphi))\psi_p + (Y\varphi)_p X_p(\psi) + X_p(\varphi)(Y\psi)_p \\
&\quad + \varphi_p X_p(Y\psi) - Y_p(X\varphi)\psi_p - (X\varphi)_p Y_p(\psi) \\
&\quad - Y_p(\varphi)(X\psi)_p - \varphi_p Y_p(X\psi) \\
&= (X_p(Y\varphi))\psi_p + \varphi_p X_p(Y\psi) - Y_p(X\varphi)\psi_p - \varphi_p Y_p(X\psi) \\
&= [X, Y]_p(\varphi)\psi_p + \varphi_p[X, Y]_p(\psi),
\end{aligned}$$

where we denote the evaluation at p with the lower index p throughout. This computation shows, that $[X, Y]_p$ satisfies the product rule (2.7), and that, therefore, $[X, Y]_p$ is a tangent vector at p in the sense of Proposition 2.2.9, and thus that $[X, Y]$ is a vector field on M.

---

[18] Henri Poincaré (1854–1912), Heinz Hopf (1894–1971).
[19] Leonhard Euler (1707–1783).

Proposition 2.4.4 *Let $X, Y \in \mathcal{V}(M)$ and let (x, U) be a chart on M. Let $\xi, \eta, \zeta : U \longrightarrow \mathbb{R}^m$ be the principal parts of X, Y and $[X, Y]$ with respect to (x, U). Then*

$$\zeta^j = \xi^i \frac{\partial \eta^j}{\partial x^i} - \eta^i \frac{\partial \xi^j}{\partial x^i}.$$

In particular, $[X, Y]$ is a smooth vector field.

We can also write the formula for the principal part of $[X, Y]$ from Proposition 2.4.4 in the following way:

$$\zeta = d\eta(X) - d\xi(Y) = X(\eta) - Y(\xi). \tag{2.22}$$

Proof of Proposition 2.4.4

From

$$\xi^j = X(x^j), \quad \eta^j = Y(x^j), \quad \zeta^j = [X, Y](x^j)$$

we obtain

$$\zeta^j = [X, Y](x^j) = X(Y(x^j)) - Y(X(x^j))$$

$$= X(\eta^j) - Y(\xi^j) = \xi^i \frac{\partial \eta^j}{\partial x^i} - \eta^i \frac{\partial \xi^j}{\partial x^i}. \qquad \square$$

Now let $f : M \longrightarrow N$ be a smooth map. Vector fields $X \in \mathcal{V}(M)$ and $Y \in \mathcal{V}(N)$ are called *f-related* if

$$f_* \circ X = Y \circ f, \text{ that is, } f_{*p}(X_p) = Y_{f(p)} \tag{2.23}$$

for all $p \in M$. For $\varphi \in \mathcal{F}(N)$ it then follows that

$$Y_{f(p)}(\varphi) = (f_{*p}(X_p))(\varphi) = X_p(\varphi \circ f),$$

so

$$(Y\varphi) \circ f = X(\varphi \circ f). \tag{2.24}$$

The following theorem on f-related vector fields is quite useful in the computation of Lie brackets.

Proposition 2.4.5 *Let $X_1, X_2 \in \mathcal{V}(M)$ be f-related to $Y_1, Y_2 \in \mathcal{V}(N)$. Then $[X_1, X_2]$ is f-related to $[Y_1, Y_2]$,*

$$f_* \circ [X_1, X_2] = [Y_1, Y_2] \circ f.$$

Proof

From (2.24) it follows that

$$[Y_1, Y_2]_{f(p)}(\varphi) = (Y_1(f(p)))(Y_2\varphi) - (Y_2(f(p)))(Y_1\varphi)$$
$$= (X_1(p))((Y_2\varphi) \circ f) - (X_2(p))((Y_1\varphi) \circ f)$$
$$= (X_1(p))(X_2(\varphi \circ f)) - (X_2(p))(X_1(\varphi \circ f))$$
$$= [X_1, X_2]_p(\varphi \circ f)$$
$$= (f_{*p}([X_1, X_2]_p))(\varphi). \qquad \square$$

Example 2.4.6

1) With (2.21) or Proposition 2.4.4,

$$\left[\frac{\partial}{\partial x^i}, \frac{\partial}{\partial x^j}\right] = 0$$

for the coordinate vector fields of a chart (U, x) on M.

2) Let L be a submanifold of M and $i: L \longrightarrow M$ the inclusion. Let Y_1 and Y_2 be vector fields on M, whose restrictions to L are tangent to L, that is, for all $p \in L$, $Y_1(p)$ and $Y_2(p)$ are in $T_p L$. The $X_j := Y_j \circ i$ are hence i-related to the Y_j, $j = 1, 2$. For all $p \in L$, then

$$[Y_1, Y_2]_p = [Y_1, Y_2]_{i(p)} = i_{*p}([X_1, X_2]_p) = [X_1, X_2]_p.$$

As a rule, we do not give restrictions of vector fields their own names.

3) We identify $\mathbb{R}^4 \cong \mathbb{H}$ and define $I, J, K \in \mathcal{V}(\mathbb{R}^4)$ via

$$I: x \mapsto xi, \quad J: x \mapsto xj, \quad K: x \mapsto xk.$$

(Strictly speaking, these are the principal parts of the vector fields with respect to the chart id on $\mathbb{H} \cong \mathbb{R}^4$.) From (2.22), it then follows that

$$[I, J]_x = xij - xji = 2xk = 2K(x)$$

and, analogously, $[J, K] = 2I$ and $[K, I] = 2J$. The restrictions of I, J, and K to $S^3 \subseteq \mathbb{R}^4$ are tangent to S^3. The formulas for the Lie bracket of the restrictions from the previous example then hold for these vector fields on S^3.

■

2.5 Vector Bundles and Sections

There are a number of other situations in which we encounter similar structures to those of tangent bundles. To this end, let F be a vector space over $\mathbb{K} \in \{\mathbb{R}, \mathbb{C}, \mathbb{H}\}$ of dimension r over $\mathbb{K}$.

Proposition and Definition 2.5.1 *A $\mathbb{K}$-vector bundle over M with fiber F and* rank r *consists of a manifold E and a smooth map $\pi \colon E \longrightarrow M$, called the* projection, *so that the following hold:*

1. *for every $p \in M$, the fiber $E_p := \pi^{-1}(p)$ over p is a $\mathbb{K}$-vector space;*
2. *there is a covering of M by open sets U together with diffeomorphisms $t \colon E|_U = \pi^{-1}(U) \longrightarrow U \times F$, called the* trivializations, *such that t is of the form*

$$t(v) = (p, \tau_p(v)), \quad p \in U \text{ and } v \in E_p,$$

and $\tau_p \colon E_p \longrightarrow F$ is an isomorphism for all $p \in U$.

A section *of the vector bundle is a map $S \colon M \longrightarrow E$ with $\pi \circ S = \mathrm{id}_M$. Together with addition and multiplication by scalars, analogous to (2.18), the set $\mathcal{S}(E)$ of smooth sections of E is a $\mathbb{K}$-vector space and a module over $\mathcal{F}(M)$. See* ▢ *Figs. 2.6 and 2.7.* □

Vector bundles with fiber F are therefore families of vector spaces E_p, which are isomorphic to F and, in the sense of the above definition, depend smoothly on $p \in M$.

Example 2.5.2

1) The model example of a vector bundle is the *trivial bundle* $E = M \times F$ with the projection $\pi(p, v) = p$ and id as a (global) trivialization.
2) The tangent bundle is a real vector bundle with fiber $\mathbb{R}^m$ and rank m. Trivializations as in Definition 2.5.1.2 are defined analogously to the charts of TM as in (2.14):

$$t_x(v) := (p, dx(p)(v)), \quad p \in U \text{ and } v \in T_pM. \tag{2.25}$$

▢ **Fig. 2.6** A trivialization of a vector bundle

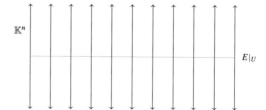

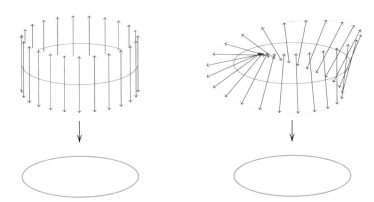

◘ Fig. 2.7 The cylinder and Möbius band as vector bundles over S^1. Up to isomorphism, these are the only 1-dimensional real vector bundles over S^1.

Conversely, for a general vector bundle, one obtains in this way *adapted charts* (with values in $\mathbb{R}^m \times F$) by choosing U with possible refinement as the domain of a chart x and replacing t in Definition 2.5.1.2 by

$$T_x(v) := (x(p), \tau_p(v)), \quad p \in U \text{ and } v \in E_p. \tag{2.26}$$

■

Construction Technique 2.5.3 Let M and F be manifolds of dimensions m and r, and let $(U_i)_{i \in I}$ be an open cover of M. For every $i, j \in I$ with $U_i \cap U_j \neq \emptyset$ let

$$f_{ij} : (U_i \cap U_j) \times F \longrightarrow (U_i \cap U_j) \times F \tag{2.27}$$

be a diffeomorphism of the form $f_{ij}(p, v) = (p, \varphi_{ij}(p, v))$, such that

$$f_{ii} = \mathrm{id} \quad \text{on } U_i \times F, \tag{2.28}$$

$$f_{jk} \circ f_{ij} = f_{ik} \quad \text{on } (U_i \cap U_j \cap U_k) \times F, \tag{2.29}$$

for all $i, j, k \in I$. Set $\tilde{E} = \cup_{i \in I} \{i\} \times U_i \times F$ and verify:

1) $(i, p, v) \sim (j, p, w)$ if $p \in U_i \cap U_j$ and $\varphi_{ij}(p, v) = w$ defines an equivalence relation on $\tilde{E}$.
2) The map $\tilde{\pi} : \tilde{E} \longrightarrow M$, $\tilde{\pi}(i, p, v) = p$, induces a map $\pi : E \longrightarrow M$, the projection, on the set $E = \{[i, p, v]\}$ of equivalence classes.
3) The maps $t_i : E|_{U_i} \longrightarrow U_i \times F$, $t_i([i, p, v]) = (p, v)$, are bijections with $(t_j \circ t_i^{-1})(p, v) = f_{ij}(p, v)$.
4) There is exactly one smooth structure on E, such that the maps t_i are diffeomorphisms. (Consider Propositions 2.4.1 and 2.4.2 and compare.) With this smooth structure, E is by

definition a *fiber bundle over M with projection π and fiber F*. Also show, that π is a submersion.

5) If F is a $\mathbb{K}$-vector space of dimension r and the $\varphi_{ij}(p,.)$ are isomorphisms of F for all $i, j \in I$ and $p \in U_i \cap U_j$, then $\pi : E \longrightarrow M$ is canonically a $\mathbb{K}$-vector bundle of rank r over M, such that the t_i are trivializations in the sense of Definition 2.5.1.

Example 2.5.4

1) With the aid of transition functions we obtain the tangent bundle TM: For charts (x, U) and (y, V) on M we define

$$f_{xy} := t_y \circ t_x^{-1} : (U \cap V) \times \mathbb{R}^m \longrightarrow (U \cap V) \times \mathbb{R}^m,$$
$$f_{xy}(p, v) = (p, D(y \circ x^{-1})|_{x(p)}(v)). \tag{2.30}$$

The conditions (2.28) and (2.29) are clearly satisfied, and the resulting vector bundle over M in the sense of the construction method of 2.5.3 is canonically isomorphic to TM. In this construction, we know the equivalence classes TM, the projection, the vector space structure on the fibers, and the trivializations (as in (2.25)) from the outset. This is the case in many examples, and helps develop intuition.

2) A second important vector bundle over M is the *cotangent bundle T^*M*. In this construction too, we know the equivalence classes T^*M, the projection, the vector space structure on the fibers, and the trivializations from the outset: The fiber T_p^*M over p is defined as the dual space of T_pM. If (U, x) is a chart on M, then $dx^i(p) \in T_p^*M$ and

$$dx^i(p)\Big(\frac{\partial}{\partial x^j}|_p\Big) = \delta_j^i, \quad \text{for all } p \in U. \tag{2.31}$$

For $p \in U$, $(dx^1(p), \ldots, dx^m(p))$ is therefore the dual basis to the basis of T_pM comprised of the vectors $\partial/\partial x^j|_p$. In particular, every $\omega \in T_p^*M$ can be uniquely written as a linear combination of the form $\omega = \omega_i dx^i(p)$ with $\omega_i = \omega(\partial/\partial x^i|_p)$. From this, we obtain a map

$$t_x : T^*M|_U \longrightarrow U \times (\mathbb{R}^m)^*, \quad t_x(\omega) = (p, \omega_i e^i), \tag{2.32}$$

with $\omega = \omega_i dx^i(p) \in T_p^*M$ and $p \in U$ as above, where (e^i) denotes the basis of $(\mathbb{R}^m)^*$ dual to the standard basis (e_i) of $\mathbb{R}^m$. Clearly, $\tau_{xp} : \omega \mapsto \omega_i e^i$ is an isomorphism for every $p \in U$. In particular, t_x is bijective. Furthermore,

$$f_{xy} := t_y \circ t_x^{-1} : (U \cap V) \times (\mathbb{R}^m)^* \longrightarrow (U \cap V) \times (\mathbb{R}^m)^*,$$
$$f_{xy}(p, w) = (p, D(x \circ y^{-1})|_{y(p)}^*(w)). \tag{2.33}$$

The reason for the reversal of the order of x and y in comparison to (2.30) is clear, as, in this example, we are discussing dual spaces and linear maps.

3) For projective spaces $M = \mathbb{K}P^n$ as in Example 2.1.15.4 let

$$E = \{(L, x) \mid x \in L \in \mathbb{K}P^n\} \quad \text{with} \quad \pi(L, x) := L.$$

In other words, the fiber over a line $L \subseteq \mathbb{K}^{n+1}$ consists of all the vectors $x \in L$, and therefore, $\pi : E \longrightarrow \mathbb{K}P^n$ is also called the *tautological bundle* over $\mathbb{K}P^n$. For $x \neq 0$, of course, $L = x\mathbb{K}$; in this sense, E therefore consists of $\mathbb{K}^{n+1} \setminus \{0\}$ and a family of zero vectors $(L, 0)$ for the $L \in \mathbb{K}P^n$. If we further identify $(L, 0)$ with L, we obtain E from $\mathbb{K}^{n+1}$ by replacing $0 \in \mathbb{K}^{n+1}$ with $\mathbb{K}P^n$. This process of replacing a point by a (suitable) projective space is called the *blow up of a point*. In our example, we blow up at the origin of $\mathbb{K}^{n+1}$.

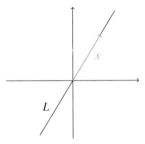

So far, we have identified the set E and the projection π. The fiber of π over $L \in \mathbb{K}P^n$ consists of the line L as a subset of $\mathbb{K}^{n+1}$ and therefore carries the structure of a one-dimensional $\mathbb{K}$-vector space.

Next, we define trivializations, and thereby obtain, as before, a smooth structure on E, such that $\pi : E \longrightarrow M$ is a $\mathbb{K}$-vector bundle of rank one. To this end, define a trivialization on the open subset $U_i \subseteq \mathbb{K}P^n$ as in Example 2.1.15.4 via

$$t_i : E|_{U_i} \longrightarrow U_i \times \mathbb{K}, \quad t_i([x], y) := ([x], y^i). \tag{2.34}$$

Because y lies in the line $[x]$, there is an $\alpha \in \mathbb{K}$ with $y = x\alpha$. On $U_i \cap U_j$, therefore

$$y^j = x^j \alpha = x^j (x^i / x^i) \alpha = (x^j / x^i) x^i \alpha = (x^j / x^i) y^i.$$

Thus

$$f_{ij}([x], y^i) := (t_j \circ t_i^{-1})([x], y^i) = ([x], (x^j / x^i) y^i).$$

The $f_{ij} = t_j \circ t_i^{-1}$ satisfy the conditions of construction 2.5.3, and therefore, the tautological bundle is a $\mathbb{K}$-vector bundle. The rank of this bundle is one, and one therefore speaks of a *line bundle*. ∎

2.6 Supplementary Literature

On can find more comprehensive introductions to the theory of manifolds in [BJ] and [Sp1, chapters 1–6]. In particular, these sources contain the interpretation of vector fields as a dynamical system or as a first-order differential equation.

An excellent, in-depth, and—nonetheless—elementary discussion of the topology of manifolds is contained in [Mi3]; [Mi1] and [Mi2] are very good sources on the theory of critical points and on cobordism theory. The three sources just mentioned are quite well-suited as templates for seminars or study groups.

2.7 Exercises

Exercise 2.7.1
For a given C^k atlas $\mathcal{A}$ on a set M there is an equivalent atlas $\mathcal{B} = ((U_i, x_i))_{i \in I}$ on M, such that $U_i' = \mathbb{R}^m$ for all $i \in I$. ■

Exercise 2.7.2
1. Compact spaces are paracompact.
2. Closed subsets of paracompact spaces are paracompact.
3. A locally connected topological spaces is paracompact if and only if its connected components are paracompact.

■

Exercise 2.7.3 (To Complete Proposition 2.1.12)
The topology of a connected manifold has a countable basis. ■

Exercise 2.7.4 (Smooth Maps)
1. For every manifold M, the identity id: $M \longrightarrow M$ is smooth. More generally: For an open subset W of M, the inclusion $W \longrightarrow M$ is smooth.
2. The composition of C^k maps is C^k.
3. A map $f : M \longrightarrow N$ is C^0 is and only if it is continuous.
4. For $f, g \in \mathcal{F}(M)$, $f + g$ and fg are also in $\mathcal{F}(M)$, where, as usual, we add and multiply functions pointwise. With these structures, $\mathcal{F}(M)$ becomes a commutative unital ring (but not a field).
5. The radial projection $\mathbb{R}^{m+1} \setminus \{0\} \longrightarrow S^m$, $x \mapsto x/\|x\|$ is smooth.
6. The canonical projection $\mathbb{K}^{n+1} \setminus \{0\} \longrightarrow \mathbb{K}P^n$, $x \mapsto [x]$, is smooth.
7. For $m \leq n$ the inclusion $\mathbb{K}P^m \hookrightarrow \mathbb{K}P^n$, $[x] \mapsto [x, 0]$, is smooth.
8. For manifolds M, N, and P, a map $f : P \longrightarrow M \times N$ is smooth if and only if $f \circ \pi_M$ and $f \circ \pi_N$ are smooth, where π_M and π_N denote the projections from $M \times N$ to M and N. Compare this with Proposition 1.4.7.

■

Exercise 2.7.5 (Diffeomorphisms)

1. Let $I \subseteq \mathbb{R}$ be an open interval. Find a diffeomorphism $f: I \longrightarrow \mathbb{R}$. The map $\mathbb{R} \longrightarrow \mathbb{R}$, $x \mapsto x^3$, is a smooth homeomorphism, but not a diffeomorphism.
2. The charts $x: U \longrightarrow U'$ on a manifold are diffeomorphisms.
3. For every orthogonal matrix $A \in \mathbb{R}^{(m+1) \times (m+1)}$, the induced map $S^m \longrightarrow S^m$, $x \mapsto Ax$, is a diffeomorphism.
4. For every invertible matrix $A \in \mathbb{K}^{(n+1) \times (n+1)}$, the induced map $\mathbb{K}P^n \longrightarrow \mathbb{K}P^n$, $[x] \mapsto [Ax]$, is a diffeomorphism.
5. Let U and V be vector spaces over $\mathbb{K}$, $A: U \longrightarrow V$ a linear map, and $0 < k < \dim_{\mathbb{K}} U, \dim_{\mathbb{K}} V$. Then the set

$$W := \{P \in G_k(U) \mid A|_P \text{ is injective}\} \subseteq G_k(U)$$

is open, and the induced map $W \longrightarrow G_k(V)$, $P \mapsto A(P)$, is smooth. If A is invertible, then this map is a diffeomorphism.
6. If M is a manifold, X a set, and $f: M \longrightarrow X$ a bijection, there is exactly one smooth structure on X (including the associated topology), such that f is a diffeomorphism. ∎

Exercise 2.7.6 (The Exponential Map on $\mathrm{Gl}(n, \mathbb{K})$)

For $A \in \mathbb{K}^{n \times n}$ set

$$\exp(A) = e^A := E + A + \frac{1}{2}A^2 + \frac{1}{3!}A^3 + \cdots,$$

where E denotes the identity matrix. Show that this series converges absolutely, and therefore that $\exp: \mathbb{K}^{n \times n} \longrightarrow \mathbb{K}^{n \times n}$ is a smooth map. Verify the following:
1. $\exp(0) = E$ and $D\exp|_0 = \mathrm{id}$;
2. $\exp(A + B) = \exp(A)\exp(B)$ if $AB = BA$.
Conclude that
a) the image of $\exp$ lies in $\mathrm{Gl}(n, \mathbb{K})$, with $\exp(-A) = \exp(A)^{-1}$ and that
b) $\mathbb{R} \longrightarrow \mathbb{K}^{n \times n}$, $t \mapsto \exp(tA)$, is smooth, with $\exp((s+t)A) = \exp(sA)\exp(tA)$. ∎

Exercise 2.7.7

The left and right translations on a Lie group G are smooth and satisfy $L_e = \mathrm{id}$, $L_{gh} = L_g \circ L_h$ and $R_{gh} = R_h \circ R_g$, where e denotes the identity element of G and g, h are arbitrary elements of G. In particular, L_g and R_g are diffeomorphisms with $(L_g)^{-1} = L_{g^{-1}}$ and $(R_g)^{-1} = R_{g^{-1}}$. ∎

Exercise 2.7.8 (Immersions)

1. For $0 < r < R$. the map $f: \mathbb{R}^2 \longrightarrow \mathbb{R}^3$,

$$f(\varphi, \psi) := ((R + r\cos\varphi)\cos\psi, (R + r\cos\varphi)\sin\psi, r\sin\varphi),$$

is an immersion. Verify that f corresponds to an embedding of the torus $T^2 = S^1 \times S^1 \longrightarrow \mathbb{R}^3$. Construct analogous embeddings $S^m \times S^n \longrightarrow \mathbb{R}^{m+n+1}$ and show that their respective images are submanifolds.

2. *Veronese embedding*[20]: Show that the map

$$\mathbb{R}P^2 \longrightarrow \mathbb{R}P^5, \quad [x, y, z] \mapsto [xx, xy, xz, yy, yz, zz],$$

is well defined and an embedding.

3. Show that the map $\mathbb{R}P^m \times \mathbb{R}P^n \longrightarrow \mathbb{R}P^{mn+m+n}$, defined by

$$([x_0, \ldots, x_m], [y_0, \ldots, y_n]) \mapsto [x_0 y_0, x_0 y_1, \ldots, x_i y_j, \ldots, x_m y_n],$$

is well defined and an embedding.

4. If M is a connected manifold, then for every two points $p \neq q$ in M there is a regular curve $c : [a, b] \longrightarrow M$ with $c(a) = p$ and $c(b) = q$.

5. For all manifolds M of dimension m and $0 < n \leq m$, there is an embedding $\mathbb{R}^n \longrightarrow M$.

6. There are injective immersions, that are not embeddings. If, however, M is compact and $f : M \longrightarrow N$ is an injective immersion, then f is an embedding.

7. For manifolds M and N and points $p \in M$ and $q \in N$,

$$i_q : M \longrightarrow M \times N, \ i_q(p') = (p', q), \quad \text{and} \quad j_p : N \longrightarrow M \times N, \ j_p(q') = (p, q'),$$

are (canonical) embeddings (sketch!), and, with respect to these,

$$T_{(p,q)}(M \times N) \cong \operatorname{im} i_{q*p} \oplus \operatorname{im} j_{p*q} \cong T_p M \oplus T_q N.$$

∎

Exercise 2.7.9 (Submersions)

1. Compute the differential of the projection

$$\pi : S^{dn-1} \longrightarrow \mathbb{K}P^{n-1}, \quad x \mapsto [x],$$

with respect to the identifications of the tangent spaces as in Examples 2.2.3.2 and 2.2.3.3 and conclude, that π is a submersion.

2. If M is compact, N connected, and $f : M \longrightarrow N$ a submersion, then f is surjective.

3. For manifolds M, N, and P and a surjective submersion $f : M \longrightarrow N$, a map $g : N \longrightarrow P$ is smooth if and only if $g \circ f$ is smooth. Compare this with Proposition 1.4.9 and Exercise 2.7.4.8.

4. If f has constant rank in a neighborhood of p, then, in a (possibly smaller) neighborhood of p, f is a composition $f = g \circ h$, where g is an embedding and h is a submersion.

∎

---

[20]Giuseppe Veronese (1854–1917).

Exercise 2.7.10 (Special Isomorphisms)

Throughout, identify $\mathbb{R}^4 \cong \mathbb{C}^2 \cong \mathbb{H}$.

0. $S^1 \subseteq \mathbb{C}^* = \mathrm{Gl}(1, \mathbb{C})$ and $S^3 \subseteq \mathbb{H}^* = \mathrm{Gl}(1, \mathbb{H})$ are Lie subgroups.

1. For $x, y \in S^3$, the map $\mathbb{R}^4 \longrightarrow \mathbb{R}^4$, $z \mapsto xz\bar{y}$, is an orthogonal transformation that preserves the orientation of $\mathbb{R}^4$. The induced map

$$f: S^3 \times S^3 \longrightarrow SO(4), \quad f(x, y)(z) := xz\bar{y},$$

   is a homomorphism, smooth with maximal rank 6, surjective, and two-to-one; in this sense, $SO(4) \cong (S^3 \times S^3)/\{\pm(1, 1)\}$.

2. For $x \in S^3$, the map $\mathbb{R}^4 \longrightarrow \mathbb{R}^4$, $z \mapsto xz\bar{x}$, is an orthogonal transformation, that leaves the imaginary quaternions $\cong \mathbb{R}^3$ invariant. The induced map

$$f: S^3 \longrightarrow SO(3), \quad f(x)(z) := xz\bar{x},$$

   is a homomorphism, smooth with maximal rank 3, surjective, and two-to-one; in this sense, $SO(3) \cong S^3/\{\pm 1\}$. Since $S^3/\{\pm 1\} \cong \mathbb{R}P^3$, therefore, $SO(3)$ and $\mathbb{R}P^3$ are diffeomorphic.

3. For $x \in S^3$, the map $\mathbb{C}^2 \longrightarrow \mathbb{C}^2$, $z \mapsto xz$, is a unitary transformation. The induced map

$$f: S^3 \longrightarrow SU(2), \quad f(x)(z) := xz,$$

   is simultaneously a group isomorphism and a diffeomorphism; in this sense, $SU(2) \cong S^3$.

∎

Exercise 2.7.11

The *Stiefel manifold*[21] $V_k(n)$ consists of the set of orthogonal k-tuples in $\mathbb{K}^n$, that is, the set of all $A \in \mathbb{K}^{n \times k}$ with $A^*A = E$. In other words, $A \in \mathbb{K}^{n \times k}$ belongs to $V_k(n)$, if the column vectors of A are orthonormal. Show that the identity matrix E is a regular value of the map $f: \mathbb{K}^{n \times k} \longrightarrow H_k(\mathbb{K})$, $f(A) := A^*A$, and therefore that $V_k(n)$ is a submanifold of $\mathbb{K}^{n \times k}$ with

$$T_A V_k(n) \cong \{B \in \mathbb{K}^{n \times k} \mid A^*B + B^*A = 0\}.$$

In the case $k = 1$, we obtain the sphere of dimension $dn - 1$, in the case $k = n$ the group G as in Example 2.3.8.3 above.

∎

Exercise 2.7.12

Find examples of (smooth or continuous) vector fields on S^2 with one zero, and examples with two zeros.

∎

---

[21] Eduard Ludwig Stiefel (1909–1978).

Exercise 2.7.13

For smooth vector fields X, Y, Z and a smooth function φ on a manifold M, the following hold:

1. $[X, Y] = -[Y, X]$;
2. $[X, \varphi Y] = (X\varphi)Y + \varphi[X, Y]$;
3. the *Jacobi identity* $[X, [Y, Z]] + [Y, [Z, X]] + [Z, [X, Y]] = 0$.

■

Exercise 2.7.14

We call a vector field X on a Lie group G *left invariant*, if $L_{g*h} X(h) = X(gh)$ for all $g, h \in G$.

1. For the general linear group $\mathrm{Gl}(n, \mathbb{K})$ and its subgroups as in Examples 2.3.8.2 and 2.3.8.3 left invariant vector fields are of the form $X_C: A \mapsto AC$. Identify the possible $C \in \mathbb{K}^{n \times n}$ for the subgroups. Further show that

$$[X_C, X_D](A) = ACD - ADC = X_{(CD-DC)}(A).$$

The Lie brackets of left invariant vector fields on these groups are themselves left invariant and correspond to the commutators of matrices in $\mathbb{K}^{n \times n}$. Also verify that the vector fields I, J, and K in Example 2.4.6.3 are left invariant on S^3 and $\mathbb{H}^* = \mathrm{Gl}(1, \mathbb{H})$.
2. Compute the left invariant vector fields on the Heisenberg group as in Example 2.1.22.3 and determine their Lie brackets.
3. Left invariant vector fields are smooth. The association $X \mapsto X(e)$ is an isomorphism from the $\mathbb{R}$-vector space of left invariant vector fields on a Lie group G and its tangent space $T_e G$ at the identity element.
4. The Lie bracket of left-invariant vector fields is left invariant.
5. Define *right invariant* vector fields and repeat the above exercises for them. Also compare the Lie brackets of left and right invariant vector fields. ■

Exercise 2.7.15

Consider the family of vector spaces $A_p^k M := A^k(T_p M)$ of k-forms as in Appendix A and construct the corresponding vector bundles $A^k M \longrightarrow M$. Note that the case $k = 1$ corresponds to the cotangent bundle. ■

Exercise 2.7.16

Discuss the tautological bundle over Grassmannian manifolds (as in Example 2.1.15.5),

$$E = \{(P, x) \mid x \in P \in G_k(V)\} \quad \text{with} \quad \pi(P, x) := P.$$

■

Exercise 2.7.17

Show that for a vector bundle E over S^1, there are connected open subsets $U_1, U_2 \subseteq S^1$ with $U_1 \cup U_2 = S^1$, such that E is trivial over U_1 and U_2, that is, admits trivializations over U_1 and U_2. (The same holds for vector bundles over S^m.) ■

Differential Forms and Cohomology

Werner Ballmann

© Springer Basel 2018
W. Ballmann, *Introduction to Geometry and Topology*, Compact Textbooks in Mathematics,
https://doi.org/10.1007/978-3-0348-0983-2_3

Differential forms play a role in various realms of mathematics. Here, we work mainly from the perspective of algebraic topology, namely, we work with de Rham cohomology.

Differential forms of degree k on a subset W of a manifold M are families ω of alternating k-linear maps $\omega(p) : (T_pM)^k \longrightarrow \mathbb{R}$, $p \in W$. Instead of differential forms of degree k, we will also speak of k-*forms*. For ease of notation, we will write $\omega_p(v_1, \ldots, v_k)$ in place of $\omega(p)(v_1, \ldots, v_k)$, depending on the situation. In the sense of Exercise 2.7.15, k-forms are sections of the vector bundle A^kM. We will not explicitly discuss this viewpoint here, but the reader may wish to keep it in mind throughout the chapter.

We have collected the prerequisites for this chapter from linear algebra in Appendices A and B. By definition, functions are differential forms of degree zero. The next case is that of differential forms of degree one, which we now discuss.

3.1 Pfaffian Forms

We also call 1-forms on $W \subseteq M$ *Pfaffian forms*.[1] A Pfaffian form ω consists of a family of linear maps $\omega(p) \colon T_pM \longrightarrow \mathbb{R}$, i.e., for all $p \in W$, $\omega(p)$ is an element of the dual space $(T_pM)^* = T_p^*M$.

By definition, Pfaffian forms are sections of the cotangent bundle T^*M, compare with Example 2.5.4.2. We call Pfaffian forms *smooth*, if they are smooth as sections of T^*M. We recall that smoothness is a local property: the restriction of a smooth Pfaffian form to an open subset U is a smooth Pfaffian form on U. If, on the other hand, ω is smooth on open sets $U_i \subseteq M$, then it is smooth on the union $W = \cup_i U_i$.

Let (U, x) be a chart on M and ω a Pfaffian form on U. Then for all $p \in U$ (Einstein summation convention!)

$$\omega_p = \omega_i(p)dx^i(p) \quad \text{with} \quad \omega_i(p) = \omega_p\left(\frac{\partial}{\partial x^i}(p)\right), \tag{3.1}$$

---

[1] Johann Friedrich Pfaff (1765–1825).

as the $dx^i(p)$ form the dual basis of the $(\partial/\partial x^j)(p)$; see also (2.31). Such formulas become more readable if we omit the point p in our notation:

$$\omega = \omega_i \, dx^i \quad \text{with} \quad \omega_i = \omega\Big(\frac{\partial}{\partial x^i}\Big). \tag{3.2}$$

The coefficients ω_i are then functions on U. By definition of the smooth structure on T^*M as in Example 2.5.4.2, ω is smooth on U if and only if the functions $\omega_i : U \longrightarrow \mathbb{R}$ are smooth.

Example 3.1.1
For all smooth functions $f : M \longrightarrow \mathbb{R}$, the differential df is a smooth Pfaffian form on M with

$$df = \frac{\partial f}{\partial x^i} dx^i \tag{3.3}$$

on the coordinate neighborhoods of charts x on M.   ∎

Definition 3.1.2

Let $W \subseteq M$ be open, and ω a smooth Pfaffian form on W. Then a smooth function $f : W \longrightarrow \mathbb{R}$ is called a *primitive function* or a *potential* of ω, if $\omega = df$.

Example 3.1.3
1) Let $W = \mathbb{R}^2 \setminus \{0\}$ and $r = r(x, y) := \sqrt{x^2 + y^2}$. Then the *winding form* ω on W is the Pfaffian form

$$\omega := \frac{1}{r^2}(x \, dy - y \, dx).$$

2) Let $W = \mathbb{R}^3 \setminus \{0\}$ and $r = r(x, y, z) := \sqrt{x^2 + y^2 + z^2}$. The Pfaffian form

$$\omega = \omega(x, y, z) := -\frac{1}{r^3}(x \, dx + y \, dy + z \, dz)$$

is called the *gravitation form* on W; $1/r$ is a potential of ω.   ∎

Let $W \subseteq M$ be open and ω be a smooth Pfaffian form on W. Additionally, let $c : [a, b] \longrightarrow W$ be a piecewise smooth curve, i.e., let there be a subdivision

$$a = t_0 < t_1 < \cdots < t_k = b, \tag{3.4}$$

such that $c|_{[t_{i-1}, t_i]}$ is smooth for all $1 \leq i \leq k$. We then set

$$\int_c \omega := \sum_{1 \leq i \leq k} \int_{t_{i-1}}^{t_i} \omega_{c(t)}(\dot{c}(t)) \, dt. \tag{3.5}$$

Now let $x: U \longrightarrow U'$ be a chart with $U \subseteq W$ and let $c([t_{i-1}, t_i]) \subseteq U$. Then $\sigma := x \circ c: [t_{i-1}, t_i] \longrightarrow U' \subseteq \mathbb{R}^m$ is smooth. With $\omega = \omega_j \, dx^j$ as in (3.2), we obtain the useful formula

$$\int_{t_{i-1}}^{t_i} \omega_{c(t)}(\dot{c}(t)) \, dt = \int_{t_{i-1}}^{t_i} \omega_j(c(t)) \cdot \dot{\sigma}^j(t) \, dt, \tag{3.6}$$

since, for all $t \in (t_{i-1}, t_i)$, $dx^j(c(t))(\dot{c}(t)) = \dot{\sigma}^j(t)$.

For smooth Pfaffian forms ω_1, ω_2 and scalars $\kappa_1, \kappa_2 \in \mathbb{R}$

$$\int_c (\kappa_1 \omega_1 + \kappa_2 \omega_2) = \kappa_1 \int_c \omega_1 + \kappa_2 \int_c \omega_2. \tag{3.7}$$

Let $\tau: [\alpha, \beta] \longrightarrow [a, b]$ be piecewise smooth and monotone with $\tau(\alpha) = a$ and $\tau(\beta) = b$ or $\tau(\alpha) = b$ and $\tau(\beta) = a$. Then $c \circ \tau$ is piecewise smooth with

$$\int_{c \circ \tau} \omega = \int_c \omega \quad \text{or} \quad \int_{c \circ \tau} \omega = -\int_c \omega \tag{3.8}$$

respectively. If $f: W \longrightarrow \mathbb{R}$ is smooth, then

$$\int_c df = f(c(b)) - f(c(a)). \tag{3.9}$$

Therefore $\int_c df$ is *path-independent*, i.e. $\int_c df$ depends only on the endpoints of c. Moreover, with respect to a chart (U, x) with $U \subseteq W$, we have

$$\frac{\partial \omega_i}{\partial x^j} = \frac{\partial \omega_j}{\partial x^i} \tag{3.10}$$

with $\omega_i = \partial f / \partial x^i$ as in (3.3), since

$$\frac{\partial \omega_i}{\partial x^j} = \frac{\partial^2 f}{\partial x^i \partial x^j} = \frac{\partial^2 f}{\partial x^j \partial x^i} = \frac{\partial \omega_j}{\partial x^i}. \tag{3.11}$$

The equations in (3.10) are therefore necessary conditions for a Pfaffian form to be the differential form of a function. In other words, the equations in (3.10) are *integrability conditions* for the equation $\omega = df$, where we are given ω and are searching for f.

3.2   Differential Forms

A *differential form of degree k* on $W \subseteq M$, or, more briefly, a *k-form* on W, is a map ω, that associates to each $p \in W$ a k-linear alternating multilinear $\omega(p): (T_p M)^k \longrightarrow \mathbb{R}$. We also write $k = \deg \omega$.

Now let $W \subseteq M$ be open and let ω be a k-form on W. Let (U, x) be a chart on M with $U \subseteq W$. Then

$$\omega = \sum_{1 \le i_1 < \cdots < i_k \le m} \omega_{i_1 \dots i_k} \, dx^{i_1} \wedge \cdots \wedge dx^{i_k} \tag{3.12}$$

on U with coefficient functions

$$\omega_{i_1 \dots i_k} = \omega^x_{i_1 \dots i_k} := \omega\left(\frac{\partial}{\partial x^{i_1}}, \dots, \frac{\partial}{\partial x^{i_k}}\right): U \longrightarrow \mathbb{R}. \tag{3.13}$$

Compare with (2.31) and Corollary A.4.

Now let (V, y) be another chart on M with $V \subseteq W$. Then on V, we have

$$\omega = \sum_{1 \le i_1 < \cdots < i_k \le m} \omega^y_{i_1 \dots i_k} \, dy_{i_1} \wedge \cdots \wedge dy_{i_k}$$

with

$$\omega^y_{i_1 \dots i_k} = \omega\left(\frac{\partial}{\partial y^{i_1}}, \dots, \frac{\partial}{\partial y^{i_k}}\right).$$

Using (2.13) and Lemma A.5, it therefore follows that

$$\begin{aligned}
\omega^y_{i_1, \dots, i_k} &= \omega\left(\frac{\partial}{\partial y^{i_1}}, \dots, \frac{\partial}{\partial y^{i_k}}\right) \\
&= \sum_{1 \le j_1 < \dots < j_k \le m} \det\left(\frac{\partial x^{j_\mu}}{\partial y^{i_\nu}}\right) \cdot \omega\left(\frac{\partial}{\partial x^{j_1}}, \dots, \frac{\partial}{\partial x^{j_k}}\right) \\
&= \sum_{1 \le j_1 < \dots < j_k \le m} \det\left(\frac{\partial x^{j_\mu}}{\partial y^{i_\nu}}\right) \cdot \omega^x_{j_1 \dots j_k}.
\end{aligned} \tag{3.14}$$

This transformation rule for the coefficients of ω is somewhat complicated. It is clear that explicit computations quickly become extensive and unpleasant. In the case $k = m$, only one determinant need be computed, which makes transformation rule (3.14) more legible,

$$\omega^y_{1 \dots m} = \det\left(\frac{\partial x^i}{\partial y^j}\right) \cdot \omega^x_{1 \dots m}. \tag{3.15}$$

We call ω *smooth*, if the coefficient functions $\omega^x_{i_1\ldots i_k}$ are smooth for all charts (U, x) on M with $U \subseteq W$. This is synonymous with the condition that ω is a smooth section of the vector bundle $A^k W$ in the sense of Exercise 2.7.15.

We now come to the *differential* of forms, more precisely to the so-called *exterior derivative*. The (exterior) derivative d associates to every smooth function (= 0-form) $f : M \longrightarrow \mathbb{R}$ the 1-form df. We would like to assign to each smooth k-form ω a smooth $(k + 1)$-form $d\omega$ as its (exterior) derivative. To this end, let $W \subseteq M$ be open, and ω a smooth k-form on W. Let (U, x) be a chart on M with $U \subseteq W$. On U, as above, we write

$$\omega = \sum_{1 \leq i_1 < \cdots < i_k \leq m} \omega_{i_1\ldots i_k} dx^{i_1} \wedge \cdots \wedge dx^{i_k} \tag{3.16}$$

and define $d\omega$ (initially only on U) via

$$d\omega := \sum_{1 \leq i_1 < \cdots < i_k \leq m} d\omega_{i_1\ldots i_k} \wedge dx^{i_1} \wedge \cdots \wedge dx^{i_k}. \tag{3.17}$$

For functions (that is 0-forms), f then agrees with the differential df.

Computation Rule 3.2.1 *For the differential d, the following hold:*
1. $d(a\omega + b\eta) = a d\omega + b d\eta$;
2. $d(\omega \wedge \eta) = d\omega \wedge \eta + (-1)^k \omega \wedge d\eta$ with $k = \deg \omega$;
3. $dd\omega = 0$.

Proof
Claim 1 follows from the linearity of the map $f \mapsto df$. For the proof of 2., we can, by 1., assume that

$$\omega = f \cdot \underbrace{dx^{i_1} \wedge \cdots \wedge dx^{i_k}}_{=:dx^I}, \quad \eta = g \cdot \underbrace{dx^{j_1} \wedge \cdots \wedge dx^{j_l}}_{=:dx^J}.$$

Then

$$d(\omega \wedge \eta) = d(f \cdot g) \wedge dx^I \wedge dx^J$$
$$= g \cdot df \wedge dx^I \wedge dx^J + f \cdot dg \wedge dx^I \wedge dx^J$$
$$= d\omega \wedge \eta + (-1)^k f \cdot dx^I \wedge dg \wedge dx^J$$
$$= d\omega \wedge \eta + (-1)^k \omega \wedge d\eta.$$

From this, 2. follows. Finally, 3. follows by calculating with ω as above:

$$d(d\omega) = d(df \wedge dx^I) = ddf \wedge dx^I - df \wedge \underbrace{ddx^I}_{=0}.$$

By the symmetry of the second partial derivative,

$$\frac{\partial^2 f}{\partial x^i \partial x^j} dx^i \wedge dx^j = -\frac{\partial^2 f}{\partial x^j \partial x^i} dx^j \wedge dx^i$$

and therefore

$$ddf = \sum_{i,j} \frac{\partial^2 f}{\partial x^i \partial x^j} dx^i \wedge dx^j = 0.$$

$\square$

It remains to show that the differential is well-defined, that is, that it is independent of the chosen chart. To this end, we derive a formula that does not involve the charts.

Proposition 3.2.2 *Let $W \subseteq M$ be open and ω a smooth k-form on W. Then for smooth vector fields $X_0, \ldots, X_k$ on W,*

$$d\omega(X_0, \ldots, X_k) = \sum_{0 \leq i \leq k} (-1)^i X_i \big(\omega(X_0, \ldots, \hat{X}_i, \ldots, X_k) \big)$$

$$+ \sum_{0 \leq i < j \leq k} (-1)^{i+j} \omega([X_i, X_j], X_0, \ldots, \hat{X}_i, \ldots, \hat{X}_j, \ldots, X_k).$$

Here, the hat over a variable indicates that it is omitted.

Proof
We denote the right hand side by $\eta = \eta(X_0, \ldots, X_k)$. Clearly, η is additive in each of the variables $X_0, \ldots, X_k$. Now let f be a smooth function on W and $i \in \{0, \ldots, k\}$ be fixed. Then

$$\eta(X_0, \ldots, f X_i, \ldots, X_k) = f \cdot \eta(X_0, \ldots, X_k)$$

$$+ \sum_{j \neq i} (-1)^j X_j(f) \cdot \omega(X_0, \ldots, \hat{X}_j, \ldots, X_k)$$

$$+ \sum_{j < i} (-1)^{i+j} X_j(f) \cdot \omega(X_i, X_0, \ldots, \hat{X}_j, \ldots, \hat{X}_i, \ldots, X_k)$$

$$- \sum_{j > i} (-1)^{i+j} X_j(f) \cdot \omega(X_i, X_0, \ldots, \hat{X}_i, \ldots, \hat{X}_j, \ldots, X_k).$$

The last three terms on the right cancel each other out. From this it follows that $\eta = \eta(X_0, \ldots, X_k)$ is homogeneous over $\mathcal{F}(W)$ in each variable.

Now let (U, x) be a chart on M with $U \subseteq W$. Since the right hand side of the desired equation is additive in ω, we can assume that ω is of the form $\omega = f dx^{i_1} \wedge \cdots \wedge dx^{i_k}$ on U.

By the properties just proved, we need only now to check that

$$\eta\left(\frac{\partial}{\partial x^{j_0}}, \ldots, \frac{\partial}{\partial x^{j_k}}\right) = (df \wedge dx^{i_1} \wedge \cdots \wedge dx^{i_k})\left(\frac{\partial}{\partial x^{j_0}}, \ldots, \frac{\partial}{\partial x^{j_k}}\right)$$

for all $1 \leq j_0 < \ldots < j_k \leq m$. Since $[\partial/\partial x^i, \partial/\partial x^j] = 0$, the sum with the Lie bracket in the definition of η vanishes. Moreover, both sides vanish if $\{i_1, \ldots, i_k\}$ is not contained in $\{j_0, \ldots, j_k\}$. If, on the other hand, $\{i_1, \ldots, i_k\} = \{j_0, \ldots, j_k\} \setminus \{j_\ell\}$ then

$$\eta\left(\frac{\partial}{\partial x^{j_0}}, \ldots, \frac{\partial}{\partial x^{j_k}}\right) = (-1)^\ell \frac{\partial f}{\partial x^{j_\ell}}$$

$$= \left(\frac{\partial f}{\partial x^{j_\ell}} dx^{j_\ell} \wedge dx^{i_1} \wedge \cdots \wedge dx^{i_k}\right)\left(\frac{\partial}{\partial x^{j_0}}, \ldots, \frac{\partial}{\partial x^{j_k}}\right)$$

$$= \left(df \wedge dx^{i_1} \wedge \cdots \wedge dx^{i_k}\right)\left(\frac{\partial}{\partial x^{j_0}}, \ldots, \frac{\partial}{\partial x^{j_k}}\right)$$

$$= d\omega\left(\frac{\partial}{\partial x^{j_0}}, \ldots, \frac{\partial}{\partial x^{j_k}}\right). \qquad \square$$

Definition 3.2.3

The $(k + 1)$-form $d\omega$ is called the *differential* of the k-form ω.

Let M and N be manifolds of dimension m and n respectively and f be a smooth map from M to N. Every k-form ω on N can be *pulled back* via f to a k-form $f^*\omega$ on M, compare with Appendix A and, in particular, (A.3):

$$(f^*\omega)_p = (f_{*p})^*\omega_{f(p)}. \tag{3.18}$$

For 0-forms h this means $f^*h = h \circ f$.

Computation Rule 3.2.4 *The pullback satisfies the following rules:*
1. $f^*(a\omega + b\eta) = af^*\omega + bf^*\eta$;
2. $f^*(\omega \wedge \eta) = f^*\omega \wedge f^*\eta$;
3. $(g \circ f)^* = f^* \circ g^*$;
4. *if ω is smooth, then $f^*\omega$ is as well, and $d(f^*\omega) = f^*(d\omega)$.*

Proof
1.–3. are exercises. For 4: let (U, x) and (V, y) be charts on M and N. Then, on $U \cap f^{-1}(V)$,

$$f^*dy^i = dy^i \circ f_* = d(y^i \circ f) = df^i \tag{3.19}$$

with $f^i := y^i \circ f$. By 1. we can assume that $\omega = h \cdot dy_{i_1} \wedge \cdots \wedge dy_{i_k}$ for a smooth function h. With 2. we obtain

$$f^*\omega = (h \circ f) \cdot df^{i_1} \wedge \cdots \wedge df^{i_k}. \tag{3.20}$$

Now $h \circ f$ and

$$df^i = \frac{\partial f^i}{\partial x^j} dx^j \tag{3.21}$$

are smooth, and thus, so is $f^*\omega$. From (3.17) and (3.20) together with $d^2 = 0$ it further follows that

$$
\begin{aligned}
d(f^*\omega) &= d(h \circ f) \wedge df^{i_1} \wedge \cdots \wedge df^{i_k} \\
&= (dh \circ f_*) \wedge df^{i_1} \wedge \cdots \wedge df^{i_k} \\
&= f^*dh \wedge f^*dy^{i_1} \wedge \cdots \wedge f^*dy^{i_k} \\
&= f^*(dh \wedge dy^{i_1} \wedge \cdots \wedge dy^{i_k}) = f^*d\omega. \qquad \square
\end{aligned}
$$

In addition to the definition of the pullback in (A.3), (3.20) and (3.21) are useful for the explicit computation of pullbacks of forms.

3.3   De Rham Cohomology

With $\mathcal{A}^k(M)$ we denote the $\mathbb{R}$-vector space and $\mathcal{F}(M)$-module of smooth k-forms on M (that is, the smooth sections of $A^k M$ as in Exercise 2.7.15). We obtain a sequence

$$\cdots \longrightarrow \mathcal{A}^k(M) \xrightarrow{\;d\;} \mathcal{A}^{k+1}(M) \xrightarrow{\;d\;} \mathcal{A}^{k+2}(M) \longrightarrow \cdots \tag{3.22}$$

of linear maps, where each composition $d^2 = d \circ d = 0$. That is, the sequence is a cochain complex as in Appendix B. The beginning is

$$\{0\} \longrightarrow \mathcal{A}^0(M) = \mathcal{F}(M) \xrightarrow{\;d\;} \mathcal{A}^1(M) \xrightarrow{\;d\;} \mathcal{A}^2(M) \longrightarrow \cdots \tag{3.23}$$

In other words: $\mathcal{A}^k(M) := \{0\}$ for all $k < 0$.

A differential form ω is called *closed* if $d\omega = 0$ and *exact* if there is a form η with $\omega = d\eta$; exactness implies closedness, since $d^2 = 0$. Two closed forms are called *cohomologous* if their difference is exact. Closed and exact forms correspond to the cocycles and coboundaries from Appendix B. We denote by $Z^k(M) \subseteq \mathcal{A}^k(M)$ and $B^k(M) \subseteq Z^k(M)$ the $\mathbb{R}$-vector subspaces of cocycles and coboundaries respectively.

Definition 3.3.1

The quotient $H^k(M) := Z^k(M)/B^k(M)$ is called the k-th *de Rham cohomology*[2] of M. The elements of $H^k(M)$ are called *de Rham cohomology classes* and the dimension $b_k(M) := \dim H^k(M)$ is called the *k-th Betti number*[3] of M.

A necessary condition for the partial differential equation $\omega = d\eta$ for given ω to have a solution η is $d\omega = 0$. For arbitrary $\omega \in Z^k(M)$ this condition is sufficient if and only if $b_k(M) = 0$.

Example 3.3.2

1) $H^1(\mathbb{R}^2) = \{0\}$: Let $\omega = f\,dx + g\,dy$ be a smooth closed 1-form. Then f and g are smooth with

$$\frac{\partial f}{\partial y} = \frac{\partial g}{\partial x}.$$

It remains to show that there is a function $h: \mathbb{R}^2 \longrightarrow \mathbb{R}$ with $dh = \omega$. Define h by

$$h(x, y) = \int_0^x f(t, 0)\,dt + \int_0^y g(x, t)\,dt.$$

Then

$$\frac{\partial h}{\partial x}(x, y) = f(x, 0) + \int_0^y \frac{\partial g}{\partial x}(x, t)\,dt$$

$$= f(x, 0) + \int_0^y \frac{\partial f}{\partial y}(x, t)\,dt$$

$$= f(x, 0) + f(x, t)\big|_{t=0}^{t=y} = f(x, y).$$

Analogously, one shows $\partial h/\partial y = g$. With this computation, we find that $dh = \omega$.

2) $H^1(\mathbb{R}^2 \setminus \{0\}) \neq 0$: The winding form

$$\omega(x, y) = \frac{1}{x^2 + y^2}(-y\,dx + x\,dy)$$

is smooth and closed, but not exact. Compare with Exercise 3.9.3.

∎

---

[2]Georges de Rham (1903–1990).
[3]Enrico Betti (1823–1892).

Proposition and Definition 3.3.3 *The wedge product on differential forms induces a product*

$$H^k(M) \times H^l(M) \longrightarrow H^{k+l}(M), \quad [\omega] \wedge [\eta] := [\omega \wedge \eta],$$

on the de Rham cohomology classes, which we also call the wedge product. *With this product,*
$H^*(M) := \oplus_{k \geq 0} H^k(M)$ *becomes an associative, graded-commutative algebra.*

Proof
The claims follow directly from the computation rules 3.2.1.   □

Proposition and Definition 3.3.4 *A smooth map $f : M \longrightarrow N$ between manifolds induces homomorphisms $H^k(N) \longrightarrow H^k(M)$ via $\omega \mapsto f^*\omega$. We also denote these homomorphisms by f^*. Moreover,*
1. $(g \circ f)^* = f^* \circ g^*$ *and* id_M *induces the identity on* $H^k(M)$.
2. $f^*([\omega] \wedge [\eta]) = f^*[\omega] \wedge f^*[\eta]$.
In particular, $f^* : H^*(N) \longrightarrow H^*(M)$ is an isomorphism of graded algebras if f is a diffeomorphism.

Proof
We leave the proof of the first claim as an exercise; also compare with Proposition B.4. For the last claim: if f is a diffeomorphism and $g := f^{-1}$, then $f^* \circ g^* = (g \circ f)^* = \mathrm{id}_{H^*(M)}$ by 1. and analogously $g^* \circ f^* = \mathrm{id}_{H^*(N)}$. Therefore g^* is inverse to f^*.   □

Corollary 3.3.5 $\mathbb{R}^2$ *and* $\mathbb{R}^2 \setminus \{0\}$ *are not diffeomorphic.*   □

Further elementary results on de Rham cohomology will be discussed in Exercise 3.9.9.

3.4   The Poincaré Lemma

The product $M \times \mathbb{R}$ is a manifold of dimension $m + 1$. We write points in $M \times \mathbb{R}$ as pairs (p, t). For $t \in \mathbb{R}$, let $i_t : M \longrightarrow M \times \mathbb{R}$ be defined by $i_t(p) := (p, t)$. If we identify $T_{(p,t)}(M \times \mathbb{R})$ with $T_p M \oplus T_t \mathbb{R} = T_p M \oplus \mathbb{R}$ as usual (compare with Exercise 2.7.8.7), then

$$(i_t)_{*p} v = (v, 0) \in T_{(p,t)}(M \times \mathbb{R}) \quad \text{for all } v \in T_p M. \tag{3.24}$$

We further set

$$\frac{\partial}{\partial t}\Big|_{(p,t)} := [s \mapsto (p, t + s)]. \tag{3.25}$$

Then $\partial/\partial t$ is a smooth vector field on $M \times \mathbb{R}$.

In the following, we consider the closed subset $M \times [0, 1]$ of $M \times \mathbb{R}$. We call a differential form ω on $M \times [0, 1]$ *smooth*, if, for every $(p, t) \in M \times [0, 1]$, there is an open neighborhood U of (p, t) in $M \times \mathbb{R}$ and a smooth differential form α on U, such that $\alpha = \omega$ on the intersection $U \cap (M \times [0, 1])$.

Lemma 3.4.1 *Let $k \geq 1$ and ω be a smooth k-form on $M \times [0, 1]$. Then there are unique smooth forms η and ζ on $M \times [0, 1]$ of degrees $k - 1$ and k respectively, such that $\omega = dt \wedge \eta + \zeta$ and*

$$\eta(v_1, \ldots, v_{k-1}) = 0 \quad and \quad \zeta(v_1, \ldots, v_k) = 0,$$

if at least one v_i is a multiple of $\partial/\partial t$.

Proof
Testing against tuples of vectors $(\partial/\partial t, v_2, \ldots, v_k)$ shows the uniqueness of η and therefore also that of ζ:

$$\omega(\partial/\partial t, v_2, \ldots, v_k) = (dt \wedge \eta)(\partial/\partial t, v_2, \ldots, v_k) + \zeta(\partial/\partial t, v_2, \ldots, v_k)$$

$$= \eta(v_2, \ldots, v_k).$$

Now let (U, x) be a chart on M. Then $(U \times \mathbb{R}, x \times \mathrm{id})$ is a chart on $M \times \mathbb{R}$, and, on $U \times [0, 1]$, we can write ω as

$$\omega = \sum_{1 \leq j_1 < \cdots < j_{k-1} \leq m} \eta_{j_1 \ldots j_{k-1}} dt \wedge dx^{j_1} \wedge \cdots \wedge dx^{j_{k-1}} + \sum_{1 \leq i_1 < \cdots < i_k \leq m} \zeta_{i_1 \ldots i_k} dx^{i_1} \wedge \cdots \wedge dx^{i_k}.$$

Over $U \times [0, 1]$, therefore,

$$\eta := \sum_{1 \leq j_1 < \cdots < j_{k-1} \leq m} \eta_{j_1 \ldots j_{k-1}} dx^{j_1} \wedge \cdots \wedge dx^{j_{k-1}} \tag{3.26}$$

and

$$\zeta := \sum_{1 \leq i_1 < \cdots < i_k \leq m} \zeta_{i_1 \ldots i_k} dx^{i_1} \wedge \cdots \wedge dx^{i_k} \tag{3.27}$$

satisfy the desired properties. The uniqueness of the representation shows that η and ζ do not depend on the choice of chart (U, x), and so are well-defined on $M \times [0, 1]$. □

Let $k \geq 1$ and ω be a smooth k-form on $M \times [0, 1]$. Write $\omega = dt \wedge \eta + \zeta$ as in Lemma 3.4.1. Define a $(k-1)$-form $I\omega$ on M via

$$(I\omega)_p(v_1, \ldots, v_{k-1}) = \int_0^1 \eta_{(p,t)}((i_t)_{*p}v_1, \ldots, (i_t)_{*p}v_{k-1}) \, dt. \tag{3.28}$$

With respect to a chart (U, x) on M and the associated chart $(U \times \mathbb{R}, x \times \mathrm{id})$ on $M \times \mathbb{R}$, we have

$$I\omega = \sum_{1 \le j_1 < \cdots < j_{k-1} \le m} \bar{\omega}_{j_1 \ldots j_{k-1}} dx^{j_1} \wedge \cdots \wedge dx^{j_{k-1}}$$

with coefficient functions

$$\bar{\omega}_{j_1 \ldots j_{k-1}}(p) = \int_0^1 \eta_{j_1 \ldots j_{k-1}}(p, t)\, dt, \tag{3.29}$$

where the $\eta_{j_1 \ldots j_{k-1}}$ are the corresponding coefficient functions of η as in (3.26). The $\bar{\omega}_{j_1 \ldots j_{k-1}}$ are therefore smooth functions on U. In particular, $I\omega$ is a smooth $(k-1)$-form on M. We thereby obtain, for all $k \ge 1$, an $\mathbb{R}$-linear operator

$$I : \mathcal{A}^k(M \times [0, 1]) \longrightarrow \mathcal{A}^{k-1}(M). \tag{3.30}$$

The operator I has the following noteworthy property.

Proposition 3.4.2 *For all smooth k-forms ω on $M \times [0, 1]$, $d(I\omega) + I(d\omega) = i_1^* \omega - i_0^* \omega$, where we set $d \circ I := 0$ on $\mathcal{A}^0(M \times [0, 1])$.*

Proof
We check this with the aid of a chart (U, x) on M. The operator I is linear in ω, and we can therefore confine ourselves to the two following cases.
1) $\omega = f dx^{i_1} \wedge \cdots \wedge dx^{i_k}$: Then $I\omega = 0$ and

$$d\omega = \frac{\partial f}{\partial t} dt \wedge dx^{i_1} \wedge \cdots \wedge dx^{i_k} + \cdots,$$

where the rest of the terms do not contain dt. For all $p \in U$, therefore,

$$(I d\omega)_p = \left(\int_0^1 \frac{\partial f}{\partial t}(p, t) dt \right) dx^{i_1} \wedge \cdots \wedge dx^{i_k}$$
$$= \left(f(p, 1) - f(p, 0) \right) dx^{i_1} \wedge \cdots \wedge dx^{i_k} = (i_1^* \omega)_p - (i_0^* \omega)_p.$$

2) $\omega = f dt \wedge dx^{j_1} \wedge \cdots \wedge dx^{j_{k-1}}$: Then

$$d\omega = \frac{\partial f}{\partial x^j} dx^j \wedge dt \wedge dx^{j_1} \wedge \cdots \wedge dx^{j_{k-1}}$$
$$= -\frac{\partial f}{\partial x^j} dt \wedge dx^j \wedge dx^{j_1} \wedge \cdots \wedge dx^{j_{k-1}},$$

and thus

$$(Id\omega)_p = -\Big(\int_0^1 \frac{\partial f}{\partial x^j}(p,t)\,dt\Big)dx^j \wedge dx^{j_1} \wedge \cdots \wedge dx^{j_{k-1}}.$$

Furthermore,

$$(dI\omega)_p = d\Big(\int_0^1 f(p,t)\,dt\Big)dx^{j_1} \wedge \cdots \wedge dx^{j_{k-1}}$$

$$= \frac{\partial}{\partial x^j}\Big(\int_0^1 f(p,t)\,dt\Big)dx^j \wedge dx^{j_1} \wedge \cdots \wedge dx^{j_{k-1}}$$

$$= \Big(\int_0^1 \frac{\partial f}{\partial x^j}(p,t)\,dt\Big)dx^j \wedge dx^{j_1} \wedge \cdots \wedge dx^{j_{k-1}}.$$

And so $dI\omega = -Id\omega$. By (3.24), on the other hand, it follows that $i_t^*(dt) = 0$ and therefore also that $i_t^*\omega = 0$. □

We can also write the formula from Proposition 3.4.2 more compactly as

$$dI + Id = i_1^* - i_0^* \tag{3.31}$$

and notice that the convention $dI = 0$ on $\mathcal{A}^0(M \times [0,1])$ is consistent with the left hand side of the sequence (3.23). We now deal with the important consequences of (3.31).

Corollary 3.4.3 *If $d\omega = 0$, then $i_1^*\omega - i_0^*\omega = d(I\omega)$.* □

Two smooth maps $f_0, f_1 \colon M \longrightarrow N$ are called *(smoothly) homotopic*, if there is a smooth map $H \colon M \times [0,1] \longrightarrow N$ such that $f_0(p) = H(p,0)$ and $f_1(p) = H(p,1)$ for all $p \in M$. Such a map H is also called a *homotopy* from f_0 to f_1.

Corollary 3.4.4 *Let $f_0, f_1 \colon M \longrightarrow N$ be homotopic smooth maps and ω a closed k-form on N, where $k \geq 0$. Then $f_0^*\omega - f_1^*\omega$ is exact. In particular, $f_0^* = f_1^* \colon H^*(N) \longrightarrow H^*(M)$.*

Proof

Let H be a homotopy from f_0 to f_1. Then $f_0 = H \circ i_0$ and $f_1 = H \circ i_1$. Since $d(H^*\omega) = H^*(d\omega) = 0$ and by Corollary 3.4.3 we thereby obtain

$$f_0^*\omega - f_1^*\omega = i_1^*(H^*\omega) - i_0^*(H^*\omega) = dI(H^*\omega). \qquad \square$$

A smooth map $f \colon M \longrightarrow N$ is called a *homotopy equivalence*, if there is a smooth map $g \colon N \longrightarrow M$, a so-called *homotopy inverse* to f, such that $g \circ f$ and $f \circ g$ are homotopic to the identity on M and N respectively.

Corollary 3.4.5 *If $f: M \longrightarrow N$ is a homotopy equivalence, then the induced map $f^*: H^*(N) \longrightarrow H^*(M)$ is an isomorphism of graded algebras.*

Proof

For a homotopy inverse $g: N \longrightarrow M$ of f, we have

$$f^* \circ g^* = (g \circ f)^* = \mathrm{id}_{H^*(M)} \quad \text{and} \quad g^* \circ f^* = (f \circ g)^* = \mathrm{id}_{H^*(N)}$$

by Corollary 3.4.4. Therefore $f^*: H^*(N) \longrightarrow H^*(M)$ is an isomorphism.   □

A manifold M is called *contractible*, if there is a smooth map $H: M \times [0, 1] \longrightarrow M$ and a point $p_0 \in M$ such that $H(p, 0) = p_0$ and $H(p, 1) = p$ for all $p \in M$. We also call such a map H a *contraction (onto p_0)*.

Example 3.4.6

1) $M = \mathbb{R}^m$ is contractible: $H(x, t) = tx$.
2) An (open) subset $W \subseteq \mathbb{R}^m$ is called *starlike*, if there is a point $p_0 \in W$ such that, for every $p \in W$, the path $tp + (1-t)p_0, 0 \leq t \leq 1$, from p_0 to p is contained in W. Starlike subsets are contractible: set $H(p, t) := tp + (1 - t)p_0$.   ∎

Poincaré Lemma 3.4.7 *If M is contractible and ω is a closed k-form on M with $k \geq 1$, then ω is exact. In other words, $H^k(M) = \{0\}$ for all $k \geq 1$.*

Proof

Let $H: M \times [0, 1] \longrightarrow M$ be a contraction. Then

$$i_1^*(H^*\omega) = (H \circ i_1)^*\omega = \mathrm{id}_M^* \, \omega = \omega.$$

Since $H \circ i_0$ is constant and $k \geq 1$, it follows that $i_0^*(H^*\omega) = (H \circ i_0)^*\omega = 0$. Now $d(H^*\omega) = H^*(d\omega) = 0$, so ω is exact: $\omega = i_1^*(H^*\omega) - i_0^*(H^*\omega) = d(IH^*\omega)$.   □

3.5   The Mayer-Vietoris Sequence and the Brouwer Fixed-Point Theorem

Let W_1 and W_2 be open subsets of M and $j_l: W_1 \cap W_2 \longrightarrow W_l$ and $i_l: W_l \longrightarrow W_1 \cup W_2$ be the inclusions, $l = 1, 2$. Define maps

$$i: \mathcal{A}^k(W_1 \cup W_2) \to \mathcal{A}^k(W_1) \oplus \mathcal{A}^k(W_2), \quad i(\omega) = (i_1^*\omega, i_2^*\omega), \tag{3.32}$$

and

$$j \colon \mathcal{A}^k(W_1) \oplus \mathcal{A}^k(W_2) \to \mathcal{A}^k(W_1 \cap W_2), \quad j(\eta, \zeta) = j_1^* \eta - j_2^* \zeta. \tag{3.33}$$

Proposition 3.5.1 *For all $k \geq 0$,*

$$0 \to \mathcal{A}^k(W_1 \cup W_2) \xrightarrow{i} \mathcal{A}^k(W_1) \oplus \mathcal{A}^k(W_2) \xrightarrow{j} \mathcal{A}^k(W_1 \cap W_2) \to 0$$

is a short exact sequence, *i.e., i is injective*, $\ker j = \operatorname{im} i$, *and j is surjective.*

Proof
With the aid of a partition of unity as in Lemma 2.1.19, one obtains smooth functions φ_1 and φ_2 on $W_1 \cup W_2$ with

$$\operatorname{supp} \varphi_1 \subseteq W_1, \quad \operatorname{supp} \varphi_2 \subseteq W_2, \quad \varphi_1, \varphi_2 \geq 0 \quad \text{and} \quad \varphi_1 + \varphi_2 = 1.$$

Now let $\omega \in \mathcal{A}^k(W_1 \cap W_2)$. Then by setting

$$\eta_p := \begin{cases} \varphi_2(p)\omega_p & \text{if } p \in W_1 \cap W_2, \\ 0 & \text{if } p \in W_1 \setminus \operatorname{supp} \varphi_2, \end{cases}$$

we obtain a k-form $\eta \in \mathcal{A}^k(W_1)$. Analogously, we obtain from $\varphi_1 \omega$ a $\zeta \in \mathcal{A}^k(W_2)$, such that $\eta_p + \zeta_p = \omega_p$ for all $p \in W_1 \cap W_2$. It then follows that $j(\eta, -\zeta) = \omega$, and therefore j is surjective.

Now let $(\eta, \zeta) \in \ker j$. Then $\eta_p = \zeta_p$ for all $p \in W_1 \cap W_2$, so by setting

$$\omega_p := \begin{cases} \eta_p & \text{if } p \in W_1, \\ \zeta_p & \text{if } p \in W_2, \end{cases}$$

we obtain a k-form $\omega \in \mathcal{A}^k(W_1 \cup W_2)$. By definition $i(\omega) = (\eta, \zeta)$, and therefore $\ker j \subseteq \operatorname{im} i$. The inclusion $\operatorname{im} i \subseteq \ker j$ is clear. Therefore $\ker j = \operatorname{im} i$. The injectivity of i is likewise clear. $\qquad \square$

In our situation, we obtain three cochain complexes over $\mathbb{R}$ (as in Appendix B) as follows: For $k \geq 0$ we set

$$\begin{aligned} C_1^k &:= \mathcal{A}^k(W_1 \cup W_2), \\ C_2^k &:= \mathcal{A}^k(W_1) \oplus \mathcal{A}^k(W_2) \\ C_3^k &:= \mathcal{A}^k(W_1 \cap W_2). \end{aligned} \tag{3.34}$$

For $k < 0$ we set $C_1^k = C_2^k = C_3^k := \{0\}$. The associated differentials d_1, d_2 and d_3 are the usual differential on differential forms of degree $k \geq 0$, so in the case $y = (\eta, \zeta) \in C_2^k$, we have $d_2 y := (d\eta, d\zeta)$. For $k < 0$ we necessarily have $d_l = 0$, $l = 1, 2, 3$. By Proposition 3.5.1, the three cochain complexes from (3.34) form a short exact sequence of cochain complexes over $\mathbb{R}$, and by Proposition B.7, therefore, induce a long exact sequence on cohomology:

Proposition and Definition 3.5.2 *For open subsets W_1 and W_2 of M, the maps i and j induce a long exact sequence, called the* Mayer-Vietoris sequence[4] *of the pair (W_1, W_2):*

$$0 \to H^0(W_1 \cup W_2) \xrightarrow{i^*} H^0(W_1) \oplus H^0(W_2) \xrightarrow{j^*} H^0(W_1 \cap W_2)$$

$$\xrightarrow{\delta} H^1(W_1 \cup W_2) \xrightarrow{i^*} H^1(W_1) \oplus H^1(W_2) \xrightarrow{j^*} \cdots \qquad\qquad \square$$

Example 3.5.3

One of the most beautiful applications of the Mayer-Vietoris sequence is the computation of the cohomology of the sphere S^m, $m \geq 2$. Since $\mathcal{A}^k(S^m) = \{0\}$ for $k > m$, $H^k(S^m) = \{0\}$ for all $k > m$. Moreover, for all $m \geq 1$, S^m is connected, so $H^0(S^m) \cong \mathbb{R}$ for all $m \geq 1$; see Exercise 3.9.9.1. By Exercise 3.9.9.4, $H^1(S^1) \cong \mathbb{R}$.

Let $m \geq 2$, and let $N = (1, 0, \ldots, 0)$, $S = (-1, 0, \ldots, 0)$ be the north and south poles in S^m. Then $W_1 = S^m \setminus \{N\}$ and $W_2 := S^m \setminus \{S\}$ are open subsets of S^m with $W_1 \cup W_2 = S^m$. The corresponding stereographic projections $W_1 \longrightarrow \mathbb{R}^m$ and $W_2 \longrightarrow \mathbb{R}^m$ are diffeomorphisms. Therefore, by the Poincaré Lemma 3.4.7, $H^0(W_1) \cong \mathbb{R} \cong H^0(W_2)$ and $H^k(W_1) = H^k(W_2) = \{0\}$ for all $k \neq 0$. Furthermore,

$$F\colon S^{m-1} \times (-\pi/2, \pi/2) \longrightarrow S^m \setminus \{N, S\} = W_1 \cap W_2,$$

$$F(x, \alpha) := ((\cos \alpha)x, \sin \alpha),$$

is a diffeomorphism. For all $k \geq 0$, therefore,

$$F^*\colon H^k(W_1 \cap W_2) \longrightarrow H^k(S^{m-1} \times (-\pi/2, \pi/2))$$

is an isomorphism. By Corollary 3.4.5 and Exercise 3.9.10.1, however,

$$H^k(S^{m-1} \times (-\pi/2, \pi/2)) \cong H^k(S^{m-1}).$$

Thus, we are in a position to recursively determine the cohomology of S^m. We assume that $H^k(S^{m-1}) \cong \mathbb{R}$ for $k = 0, m-1$, and that $H^k(S^{m-1}) = \{0\}$ for all other k. Next, we consider the beginning

$$0 \longrightarrow H^0(S^m) \xrightarrow{i^*} H^0(W_1) \oplus H^0(W_2) \xrightarrow{j^*} H^0(W_1 \cap W_2)$$

---

[4]Walther Mayer (1887–1948), Leopold Vietoris (1891–2002).

of the Mayer-Vietoris sequence, and recall for the following discussion that this sequence is exact. Since $m \geq 2$, $W_1 \cap W_2$ is connected, as are W_1, W_2 and $W_1 \cup W_2$. Now i^* is injective on $H^0(S^m)$, so the image of i^* is a one-dimensional subspace in $H^0(W_1) \oplus H^0(W_2) \cong \mathbb{R}^2$. Therefore, the image of j^* in $H^0(W_1 \cap W_2) \cong \mathbb{R}$ is one-dimensional, and so is equal to $H^0(W_1 \cap W_2)$, i.e. j^* is surjective. It therefore follows that $\delta = 0$ on $H^0(W_1 \cap W_2)$. So we obtain an exact sequence

$$0 \longrightarrow H^1(S^m) \longrightarrow H^1(W_1) \oplus H^1(W_2) = 0,$$

and therefore $H^1(S^m) = 0$. We further obtain exact sequences

$$0 \longrightarrow H^k(W_1 \cap W_2) \xrightarrow{\delta} H^{k+1}(S^m) \longrightarrow 0$$

for all $k \geq 1$, so δ is an isomorphism for all $k \geq 1$. Thus,

$$H^k(S^m) \cong \begin{cases} \mathbb{R} & \text{if } k = 0, m, \\ 0 & \text{otherwise.} \end{cases} \tag{3.35}$$

The Betti number $b_k(S^m)$ of S^m is therefore 1 for $k = 0, m$ and 0 otherwise. ∎

Brouwer Fixed-Point Theorem[5] 3.5.4 (Smooth Case) *Let* $m \geq 1$,

$$B^m := \{x \in \mathbb{R}^m \mid \|x\| \leq 1\}, \text{ and } f: B^m \longrightarrow B^m \text{ be smooth.}$$

Then f *has a* fixed point, *that is, a point* $x \in B^m$ *with* $f(x) = x$.

Proof

The case $m = 1$ follows from the Intermediate Value Theorem. We can therefore take $m \geq 2$. We further assume that f has no fixed point. Then, for $x \in B^m$, let $g(x)$ be the intersection point of the ray

$$tx + (1 - t)f(x), \quad t > 0,$$

with S^{m-1}. We thereby obtain a smooth map $g: B^m \longrightarrow S^{m-1}$ with $g(x) = x$ for all $x \in S^{m-1}$; see ◻ Fig. 3.1. We now consider the homotopy

$$H: S^{m-1} \times [0, 1] \longrightarrow S^{m-1}, \quad H(x, t) = g(tx).$$

On S^{m-1}, then, $H_1 = H(., 1) = g = \mathrm{id}$ and $H_0 = H(., 0) = \text{const}$. So H_1 induces the identity on $H^{m-1}(S^{m-1})$. Since H_0 is constant and $m - 1 \geq 1$, on the other hand,

---

[5] Luitzen Egbertus Jan Brouwer (1881–1966).

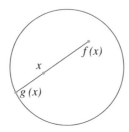

Fig. 3.1 The line from $f(x)$ to $g(x)$ through x

$H_0^* = 0$ on $H^{m-1}(S^{m-1})$. By Corollary 3.4.4, however, $H_0^* = H_1^*$, in contradiction to $H^{m-1}(S^{m-1}) \neq \{0\}$. $\qquad\square$

3.6   Orientations and the Jordan-Brouwer Theorem

Let V be an m-dimensional vector space over $\mathbb{R}$. Then bases $(b_1, \ldots, b_m)$ and $(b_1', \ldots, b_m')$ of V are called *consistently oriented*, if the automorphism of V which sends b_i to b_i', $1 \le i \le m$, has positive determinant. This defines an equivalence relation on the set of bases of V with two equivalence classes, called the *orientations* of V. If $\mathcal{O}$ is a chosen orientation of V, then we call V together with $\mathcal{O}$ an *oriented vector space*, the bases in $\mathcal{O}$ *positively oriented*, and the other bases of V *negatively oriented*. The standard basis $(e_1, \ldots, e_m)$ determines the *canonical orientation* of $\mathbb{R}^m$.

We now wish to transfer the concept of orientation to manifolds. To this end, let M be a manifold of dimension m.

Definition 3.6.1

An *orientation* of M consists of a family $\mathcal{O} = (\mathcal{O}_p)$ of orientations of the $T_p M$, $p \in M$, such that, for every $p \in M$, there is a chart (U, x) on M around p such that, for all $q \in U$,

$$\left(\frac{\partial}{\partial x^1}\Big|_q, \ldots, \frac{\partial}{\partial x^m}\Big|_q \right)$$

is a positively oriented basis of $T_q M$ with respect to $\mathcal{O}_q$. Such charts are then called *positively oriented* (with respect to $\mathcal{O}$). We call M *orientable* if M can be given an orientation, and *oriented* when an orientation on M has been chosen. We call a local diffeomorphism $f : M \longrightarrow N$ between oriented manifolds *orientation preserving*, if f_{*p} sends positively oriented bases of $T_p M$ to positively oriented bases of $T_{f(p)} N$ for all $p \in M$.

Example 3.6.2

Let V be an m-dimensional vector space over $\mathbb{R}$ and $\mathcal{O}$ an orientation of V as a vector space. Now consider V as a manifold as in Example 2.1.15.1 and call a basis of $T_v V$,

$v \in V$, positively oriented if it belongs to $\mathcal{O}$ under the usual identification $T_v V \cong V$ as in Example 2.2.3.1. The charts x_B with $B \in \mathcal{O}$ are then positively oriented, and V is an oriented manifold when equipped with these. ∎

Orientability is a generalization of two-sidedness; compare also with Remark 4.3.9. The Möbius band is one-sided and therefore not orientable. We now want to examine two-sidedness more closely.

Example 3.6.3
Let V be a real vector space of dimension $m \geq 2$, $\mathcal{O}_V$ be an orientation of V, and $f : V \longrightarrow \mathbb{R}$ be a non-trivial linear map. Then $L = f^{-1}(0)$ is an $(m-1)$-dimensional subspace of V, and the two open half-spaces $W_- := f^{-1}((-\infty, 0))$ and $W_+ := f^{-1}((0, \infty))$ of V have L as their common boundary. We call a basis $(b_2, \ldots, b_m)$ of L positively oriented if the basis $(b_1, b_2, \ldots, b_m)$ of V is positively oriented with respect to $\mathcal{O}_V$ for one (and therefore any) $b_1 \in W_+$. With this construction, L becomes an oriented vector space over $\mathbb{R}$. ∎

Now let M be a manifold of dimension $m \geq 2$ and $D \subseteq M$ be a *domain*, that is, an open subset of M. We call $p \in \partial D$ a *regular boundary point* of D, if there is a chart

$$x : U \longrightarrow U' = (-r, r) \times U''$$

on M around p with

$$x^{-1}(\{0\} \times U'') = \partial D \cap U \quad \text{and} \quad x^{-1}((-r, 0) \times U'') = D \cap U; \qquad (3.36)$$

See ◻ Fig. 3.2. We call the other boundary points of D *singular*. By $\partial_R D$ we denote the set of regular boundary points of D, and by $\partial_S D := \partial D \setminus \partial_R D$ the set of singular

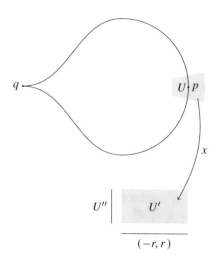

◻ **Fig. 3.2** A regular boundary point p and singular boundary point q

boundary points. The latter is closed in M, because ∂D is closed in M and $\partial_R D$ is open in ∂D.

By definition, $\partial_R D$ is an $(m-1)$-dimensional submanifold of M. Let $p \in \partial_R D$. We then say that $v \in T_p M$ is an *inward-pointing vector* (with respect to D), if every smooth curve $c = c(t)$ through p with $\dot{c}(0) = v$ lies in D for all sufficiently small values of $t > 0$. With respect to a chart x around p as in (3.36), these are precisely the vectors

$$\xi^1 \frac{\partial}{\partial x^1}\Big|_p + \cdots + \xi^m \frac{\partial}{\partial x^m}\Big|_p \quad \text{with } \xi^1 < 0.$$

We call $v \in T_p M$ an *outward-pointing vector* if $-v$ is an inward-pointing vector. We now extend Example 3.6.3 to the regular boundary components of domains.

Proposition and Definition 3.6.4 *An orientation $\mathcal{O}$ of M induces an orientation on $\partial_R D$: For $p \in \partial_R D$, we call a basis $(b_2, \ldots, b_m)$ of $T_p \partial_R D$ positively oriented, if the basis $(b_1, b_2, \ldots, b_m)$ of $T_p M$ for one (and thus any) outward-pointing vector $b_1 \in T_p M$ is positively oriented with respect to the given orientation $\mathcal{O}_p$ of $T_p M$. With this construction, $\partial_R D$ becomes an $(m-1)$-dimensional oriented manifold.*

Proof

We can assume $m \geq 2$. Now let x be a chart around $p \in \partial_R D$ as in (3.36). If x is a negatively oriented chart on M, we replace x^2 by $-x^2$ and thereby obtain a positively oriented chart. Then

$$\left(\frac{\partial}{\partial x^2}\Big|_q, \ldots, \frac{\partial}{\partial x^m}\Big|_q \right)$$

is a positively oriented basis of $T_q \partial_R D$ for all $q \in U \cap \partial_R D$. $\square$

Example 3.6.5

The sphere S^m is the boundary of the open ball $D = \{x \in \mathbb{R}^{m+1} \mid \|x\| < 1\}$. For $x \in S^m$, $T_x S^m = x^\perp$ and x is an outward-pointing vector with respect to D. A basis $(b_1, \ldots, b_m)$ of $T_x S^m$ is therefore, by Definition 3.6.4, positively oriented if $(x, b_1, \ldots, b_m)$ is a positively oriented basis of $\mathbb{R}^{m+1}$ (with respect to the canonical orientation). With this construction, S^m becomes an oriented manifold.

More generally, let M be an oriented manifold, and $f : M \longrightarrow \mathbb{R}$ a smooth function that has $a \in \mathbb{R}$ as a regular value. Then $D = f^{-1}((-\infty, a))$ is a domain with regular boundary $\partial D = \partial_R D = f^{-1}(a)$. A vector $v \in T_p M$ with $p \in \partial D$ is an outward-pointing vector if and only if $df(v) > 0$. The sphere fits into this class of examples with $M = \mathbb{R}^{m+1}$ and $f = f(x) = \|x\|^2$. ■

Now let M be a compact oriented manifold of dimension $m - 1$ and $f : M \longrightarrow S^m$ be a smooth embedding. We equip S^m with the orientation from Example 3.6.5. For $p \in M$, let $v(p)$ be the unit vector in $T_{f(p)} S^m$ perpendicular to $\mathrm{im}\, f_{*p}$, such that $(v(p), f_* b_2, \ldots, f_* b_m)$ is a positively oriented basis of $T_{f(p)} S^m = f(p)^\perp$ for all positively oriented bases $(b_2, \ldots, b_m)$ of $T_p M$.

Lemma 3.6.6 *The map* $v \colon M \longrightarrow \mathbb{R}^{m+1}$ *is smooth, and*

$$F \colon M \times (-\varepsilon, \varepsilon) \longrightarrow S^m, \quad F(p, \alpha) = \cos \alpha \cdot f(p) + \sin \alpha \cdot v(p),$$

is, for sufficiently small $\varepsilon > 0$, *a diffeomorphism onto an open neighborhood of the image* $f(M)$ *of* f *in* S^m.

Proof

Let $p \in M$. With respect to the usual identification

$$T_{(p,0)}(M \times \mathbb{R}) \cong T_p M \oplus \mathbb{R},$$

we have $F_{*(p,0)}(v, 0) = f_{*p}(v)$ and $F_{*(p,0)}(\partial/\partial \alpha) = v(p)$. Therefore $F_{*(p,0)}$ is surjective and thus, since $\dim T_p M + 1 = m$, it is an isomorphism. By the Inverse Function Theorem 2.2.13 it therefore follows that there are open neighborhoods U_p of $(p, 0)$ in $M \times \mathbb{R}$ and V_p of $f(p)$ in S^m, such that $F \colon U_p \longrightarrow V_p$ is a diffeomorphism.

It only remains to show that F is injective for sufficiently small $\varepsilon > 0$. Were this not the case, then there would be sequences (p_n, α_n) and (q_n, β_n) with $p_n, q_n \in M$ and $\alpha_n, \beta_n \in \mathbb{R}$, such that $\alpha_n, \beta_n \longrightarrow 0$, $(p_n, \alpha_n) \neq (q_n, \beta_n)$ and $F(p_n, \alpha_n) = F(q_n, \beta_n)$. Since M is compact we can assume $p_n \longrightarrow p \in M$ and $q_n \longrightarrow q \in M$ by passing to subsequences. Then $f(p) = F(p, 0) = F(q, 0) = f(q)$; so $p = q$, since f is injective. Then, however, (p_n, α_n) and (q_n, β_n) are in U_p for all sufficiently n, a contradiction. □

Jordan-Brouwer Separation Theorem 3.6.7 *Let M be compact and oriented of dimension $m - 1 \geq 1$ and $f \colon M \longrightarrow S^m$ be an embedding. Then the complement $S^m \setminus f(M)$ of the image of $f(M)$ splits into $z + 1$ connected components, where z denotes the number of connected components of M. Each connected component of the image $f(M)$ is in the boundary of precisely two of the connected components of $S^m \setminus f(M)$. See* ◘ *Fig. 3.3.*

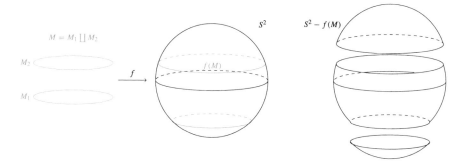

◘ **Fig. 3.3** The Jordan Brouwer Separation Theorem for $m = 2$

Proof

Let $W_1 := \operatorname{im} F$ with F and $\varepsilon > 0$ as in Lemma 3.6.6 and let $W_2 := S^m \setminus f(M)$. Both are open subsets in S^m, and $W_1 \cup W_2 = S^m$. For all $k \geq 0$, $H^k(W_1) \cong H^k(M \times (-\varepsilon, \varepsilon))$, since F is a diffeomorphism, and $H^k(M \times (-\varepsilon, \varepsilon)) \cong H^k(M)$ by Exercise 3.9.10.1. Furthermore, $F: M \times ((-\varepsilon, 0) \cup (0, \varepsilon)) \longrightarrow W_1 \cap W_2$ is a diffeomorphism, so $W_1 \cap W_2$ consists of $2z$ connected components.

We now consider the beginning of the Mayer-Vietoris sequence,

$$0 \to H^0(S^m) \overset{i^*}{\to} H^0(W_1) \oplus H^0(W_2) \overset{j^*}{\to} H^0(W_1 \cap W_2) \to H^1(S^m).$$

Since $m \geq 2$, we have $H^0(S^m) \cong \mathbb{R}$ and $H^1(S^m) = 0$. Furthermore,

$$H^0(W_1) \cong \mathbb{R}^z \quad \text{and} \quad H^0(W_1 \cap W_2) \cong \mathbb{R}^{2z}.$$

We thereby obtain a short exact sequence

$$0 \to H^0(S^m) \overset{i^*}{\to} H^0(W_1) \oplus H^0(W_2) \overset{j^*}{\to} H^0(W_1 \cap W_2) \to 0 \tag{3.37}$$

and conclude $H^0(W_2) \cong \mathbb{R}^{z+1}$. So $W_2 = S^m \setminus f(M)$ consists of $z+1$ connected components.

The connected components of $f(M)$ are the images of the connected components of M, since f is a diffeomorphism of M onto the submanifold $f(M) \subseteq S^m$. For every connected component M_i of M, $1 \leq i \leq z$, we obtain precisely two different connected components of $W_1 \cap W_2$, $U_{2i-1} := F(M_i \times (-\varepsilon, 0))$ and $U_{2i} := F(M_i \times (0, \varepsilon))$. The U_j are each contained in exactly one connected component of W_2. Let

$$f_i: W_1 \longrightarrow \mathbb{R}, \quad g_j: W_1 \cap W_2 \longrightarrow \mathbb{R}, \quad \text{and} \quad h_k: W_2 \longrightarrow \mathbb{R}$$

be the characteristic functions of the $F(M_i \times (-\varepsilon, \varepsilon))$, the U_j, and the connected components V_k of W_2, $1 \leq k \leq z + 1$ respectively. Then the $[f_1], \ldots, [f_z]$ form a basis of $H^0(W_1)$, the $[g_1], \ldots, [g_{2z}]$ a basis of $H^0(W_1 \cap W_2)$, and the $[h_1], \ldots, [h_{z+1}]$ a basis of $H^0(W_2)$. Since $f_i = g_{2i-1} + g_{2i}$ on $W_1 \cap W_2$, it follows that $j^*([f_i], 0) = [g_{2i-1}] + [g_{2i}]$. Furthermore, $j^*(0, [h_k])$ is the negative of the sum of those $[g_j]$ such that U_j is contained in V_k. This means that the set $\{1, \ldots, 2z\}$ is divided into disjoint subsets I_k, $1 \leq k \leq z + 1$, such that $(0, [h_k])$ is mapped to the negative of the sum of the $[g_j]$ with $j \in I_k$. If one of the I_k contained both the indices $2i - 1$ and $2i$ for any $1 \leq i \leq z$, then $[g_{2i-1}] - [g_{2i}] \notin \operatorname{im} j^*$. Then j^* would not be surjective, a contradiction to the exactness of (3.37). $\square$

An interesting corollary of this theorem is a special case of the Jordan Curve Theorem, namely the case where the Jordan curve $c: S^1 \longrightarrow \mathbb{R}^2$ is smooth and regular. By Exercise 1.9.17, we can compose this with the embedding of the one-point compactification $f: \mathbb{R}^2 \longrightarrow S^2$ to obtain an embedding $S^1 \longrightarrow S^2$ satisfying the hypotheses of the Jordan-Brouwer Separation Theorem. Since, by definition, this embedding misses the point ∞, we then get a decomposition of $S^2 \setminus f(c(S^1))$ into two connected components, one containing ∞, and one not. Tracing back through the definitions, we see that this implies that $\mathbb{R}^2 \setminus c(S^1)$ is divided into two connected components, one bounded, and the other unbounded.

3.7   The Oriented Integral and Stokes's Integral Formula

Let $A \subseteq M$ be Lebesgue-measurable[6] and ω be an m-form over A. Our goal is the definition of the integral $\int_A \omega$. We therefore decompose A into at most countably many Lebesgue-measurable subsets A_μ, such that the following conditions hold:
1. The A_μ are pairwise disjoint and $\cup_\mu A_\mu = A$.
2. For every μ, there is a chart (U_μ, x_μ) on M with $A_\mu \subseteq U_\mu$.

On U_μ, we then write (with $x_\mu = (x_\mu^1, \ldots, x_\mu^m)$)

$$\omega = f_\mu \, dx_\mu^1 \wedge \cdots \wedge dx_\mu^m \tag{3.38}$$

and set

$$\int_A \omega := \sum_\mu \int_{x_\mu(A_\mu)} f_\mu \circ x_\mu^{-1}, \tag{3.39}$$

where the right-hand side is integrated with respect to the usual Lebesgue measure on $\mathbb{R}^m$. It now remains for us to clarify the conditions under which the integral $\int_A \omega$ is well-defined.

Lemma 3.7.1 *Let (U, x) and (V, y) be two charts on M, such that, for all $p \in U \cap V$,*

$$\left(\frac{\partial}{\partial x^1}(p), \ldots, \frac{\partial}{\partial x^m}(p)\right) \quad \text{and} \quad \left(\frac{\partial}{\partial y^1}(p), \ldots, \frac{\partial}{\partial y^m}(p)\right)$$

are
1. consistently or
2. inversely oriented.
Let $A \subseteq U \cap V$ be Lebesgue-measurable and write

$$\omega = f \, dx^1 \wedge \cdots \wedge dx^m = g \, dy^1 \wedge \cdots \wedge dy^m$$

on $U \cap V$. Then $g \circ y^{-1}$ is integrable on $y(A)$ if $f \circ x^{-1}$ is integrable on $x(A)$. In this case

$$\int_{x(A)} f \circ x^{-1} = \pm \int_{y(A)} g \circ y^{-1},$$

where the $\pm$ is positive in case 1 and negative in case 2.

---

[6]Henri Léon Lebesgue (1875–1941).

Proof

Let $\tau \colon x(U \cap V) \longrightarrow y(U \cap V)$ be the change of coordinates, that is $\tau = y \circ x^{-1}$. For all $u \in x(U \cap V)$, $\det D\tau|_u > 0$ in case 1 and $\det D\tau|_u < 0$ in case 2. By (3.15),

$$f = g \cdot \det((\partial x_i / \partial y_j)), \quad \text{so} \quad f \circ x^{-1} = (g \circ y^{-1} \circ \tau) \cdot \det D\tau.$$

By the transformation rules for the Lebesgue integral, it then follows that $f \circ x^{-1}$ is integrable on $x(A)$ if and only if $g \circ y^{-1}$ is integrable over $y(A)$, and that then

$$\int_{x(A)} f \circ x^{-1} = \pm \int_{\tau^{-1}(y(A))} (g \circ y^{-1} \circ \tau) \cdot |\det D\tau| = \pm \int_{y(A)} g \circ y^{-1},$$

where the $\pm$ is positive in case 1 and negative in case 2. $\square$

An orientation on M over A is therefore a necessary piece of information in the formula (3.39) for the integral. The integral depends on such orientations. To ensure the integral is well-defined, we will assume that M is oriented. In the decomposition $A = \cup_\mu A_\mu$, we refine the conditions 3.7. and 3.7. defined above:

3) The (U_μ, x_μ) are positively oriented charts on M.

We call ω *integrable* over A, if

$$\sum_\mu \int_{x_\mu(A_\mu)} |f_\mu \circ x_\mu^{-1}| < \infty. \tag{3.40}$$

We now must show that the integrability of ω and the integral (3.39) do not depend on the choice of decomposition or the choice of positively oriented charts:

Lemma 3.7.2 *Let M be oriented, and $A \subseteq M$ be Lebesgue measurable. Let $A = \cup A_\mu = \cup B_\nu$ be a decomposition of A into Lebesgue measurable subsets. For A_μ and B_ν, let (U_μ, x_μ) and (V_ν, y_ν) be positively oriented charts with $A_\mu \subseteq U_\mu$ and $B_\nu \subseteq V_\nu$.*

Let ω be an m-form over A. Then ω is integrable with respect to the data (A_μ, x_μ) if and only if ω is integrable with respect to the data (B_ν, y_ν), and then

$$\sum_\mu \int_{x_\mu(A_\nu)} f_\mu \circ x_\mu^{-1} = \sum_\nu \int_{y_\nu(B_\nu)} g_\nu \circ y_\nu^{-1},$$

where the f_μ and g_ν are the coefficients of ω with respect to x_μ and y_ν in the sense of (3.38).

Proof

Let $\tau_{\nu\mu}$ be the change of coordinates, $\tau_{\nu\mu} = y_\nu \circ x_\mu^{-1}$. As in the proof of Lemma 3.7.1, we obtain

$$\sum_\mu \int_{x_\mu(A_\mu)} |f_\mu \circ x_\mu^{-1}| = \sum_{\mu,\nu} \int_{x_\mu(A_\mu \cap B_\nu)} |f_\mu \circ x_\mu^{-1}|$$

$$= \sum_{\mu,\nu} \int_{\tau_{\nu\mu}^{-1}(y_\nu(A_\mu \cap B_\nu))} |g_\nu \circ y_\nu^{-1} \circ \tau_{\nu\mu}| \cdot |\det D\tau_{\nu\mu}|$$

$$= \sum_{\mu,\nu} \int_{y_\nu(A_\mu \cap B_\nu)} |g_\nu \circ y_\nu^{-1}|$$

$$= \sum_\nu \int_{y(B_\nu)} |g_\nu \circ y_\nu^{-1}|.$$

This shows that the integrability of ω is well-defined. The independence of the integral follows from the same computation with the the absolute values omitted. In this computation, it is essential that $\det D\tau_{\mu\nu} > 0$. $\qquad\square$

Example 3.7.3

If ω is a smooth m-form with compact support, then ω is integrable. $\qquad\blacksquare$

Proposition 3.7.4 *Let $h: V \longrightarrow M$ be an orientation-preserving diffeomorphism on the open subset $W \subseteq M$ and ω be an integrable m-form on the Lebesgue measurable subset $A \subseteq W$. Then, setting $B := h^{-1}(A)$, we have*

$$\int_A \omega = \int_B h^*\omega.$$

Proof

Choose a decomposition $A = \cup A_\mu$ and positively oriented charts (U_μ, x_μ) of M as in the definition of the integral $\int_A \omega$, such that, additionally, $U_\mu \subseteq W$ for all μ. Then the (V_μ, y_μ), with $V_\mu = f^{-1}(U_\mu)$ and $y_\mu = x_\mu \circ f$, are positively oriented charts on V that cover B, such that $h^* dx_\mu^i = dy_\mu^i$. With

$$\omega = f_\mu \, dx_\mu^1 \wedge \cdots \wedge dx_\mu^m \quad \text{and} \quad h^*\omega = g_\mu \, dy_\mu^1 \wedge \cdots \wedge dy_\mu^m$$

we then have $g_\mu = f_\mu \circ h$ with respect to these special charts. Now decompose B into the $B_\mu = h^{-1}(A_\mu)$. In summary, it then follows that $x_\mu(A_\mu) = y_\mu(B_\mu)$ and $f_\mu \circ x_\mu^{-1} = g_\mu \circ y_\mu^{-1}$. $\qquad\square$

A (paraxial) compact or open *cuboid* in R^m is a subset of the form $I_1 \times \cdots \times I_m$, where the I_j are respectively compact or open intervals.

Stokes's Integral Formula for cuboids[7] 3.7.5 *Let* $Q \subseteq \mathbb{R}^m$ *be a compact cuboid,* V *be an open neighborhood of* Q *in* $\mathbb{R}^m$, *and* $h \colon V \longrightarrow M$ *be an orientation-preserving diffeomorphism to the open subset* $W \subseteq M$. *Then if* ω *is a smooth* $(m-1)$-*form on* W

$$\int_P d\omega = \int_{\partial_R P} \omega,$$

where $\partial_R P$ *is the regular part of the boundary of* $P = h(Q)$ *equipped with the induced orientation as in Proposition 3.6.4.*

Proof
We pull ω back along h to $Q = [a_1, b_1] \times \cdots \times [a_m, b_m]$. The regular part of the boundary of Q consists of the pieces

$$Q(x_i) := (a_1, b_1) \times \ldots (a_{i-1}, b_{i-1}) \times \{x_i\} \times (a_{i-1}, b_{i-1}) \times \cdots \times (a_m, b_m)$$

with $x_i = a_i$ or $x_i = b_i$. The regular part of the boundary of P consists of the images of these sets under h, since h is a diffeomorphism (on the neighborhood V of Q). Since h is orientation-preserving and $dh^*\omega = h^*d\omega$, it follows that $\int_P d\omega = \int_Q dh^*\omega$. With respect to the induced orientation as in Proposition 3.6.4, it further holds that

$$\int_{h(Q(a_i))} \omega = (-1)^i \int_{Q(a_i)} h^*\omega \quad \text{and} \quad \int_{h(Q(b_i))} \omega = (-1)^{i-1} \int_{Q(b_i)} h^*\omega,$$

where we view $Q(a_i)$ and $Q(b_i)$ each as canonically oriented open cuboids in $\mathbb{R}^{m-1}$. With appropriate smooth functions f_j we now have

$$h^*\omega = \sum_{1 \leq i \leq m} f_i \, dx^1 \wedge \cdots \wedge \widehat{dx^i} \wedge \cdots \wedge dx^m.$$

Therefore,

$$dh^*\omega = \sum_{1 \leq i, j \leq m} \frac{\partial f_i}{\partial x^j} \, dx^j \wedge dx^1 \wedge \cdots \wedge \widehat{dx^i} \wedge \cdots \wedge dx^m$$

$$= \sum_{1 \leq i \leq m} (-1)^{i-1} \frac{\partial f_i}{\partial x^i} \, dx^1 \wedge \cdots \wedge dx^m.$$

---

[7] Sir George Gabriel Stokes (1819–1903).

Using the Fubini Theorem[8] and the Fundamental Theorem of differential and integral calculus, we obtain

$$\int_Q \frac{\partial f_i}{\partial x^i}\, dx^1 \ldots dx^m = \int_{Q_i} \int_{[a_i, b_i]} \frac{\partial f_i}{\partial x^i}\, dx^i dx^1 \ldots \widehat{dx^i} \ldots dx^m$$

$$= \int_{Q_i} (f_i(x', b_i, x'') - f_i(x', a_i, x''))dx^1 \ldots \widehat{dx^i} \ldots dx^m,$$

where $x' = (x^1, \ldots, x^{i-1})$, $x'' = (x^{i+1}, \ldots, x^m)$, and $Q_i \subseteq \mathbb{R}^{m-1}$ is the product of the intervals $[a_j, b_j]$ with $j \neq i$. However, the term on right-hand side is

$$\int_{Q(b_i)} h^*\omega - \int_{Q(a_i)} h^*\omega,$$

since the terms of $h^*\omega$ above which contain dx^i vanish on the $Q(x_i)$. With respect to the orientations of $Q(a_i)$ and $Q(b_i)$ as boundary components of Q, we get a factor $(-1)^{i-1}$. The claim follows from this. □

Stokes's Integral Formula 3.7.6 (Smooth Case) *Let M be an oriented manifold of dimension m and $D \subseteq M$ a domain with smooth boundary, that is, with $\partial_S D = \emptyset$. Let ω be a smooth $(m - 1)$-form on M with compact support. Then*

$$\int_D d\omega = \int_{\partial D} \omega.$$

Proof
For every $p \in D$, there is a chart $x_p: U_p \longrightarrow (-2, 2)^m$ around p with $U_p \subseteq D$. For every $p \in \partial D$, there is a chart $x_p: U_p \longrightarrow (-2, 2)^m$ around p with $x_p(\partial D \cap U) = \{0\} \times (-2, 2)^{m-1}$ and $x_p(D \cap U) = (-2, 0) \times (-2, 2)^{m-1}$. Since the support $\operatorname{supp} \omega$ of ω is compact, there are finitely many points $p_1, \ldots, p_n$ in $\bar{D}$, such that

$$\bar{D} \cap \operatorname{supp} \omega \subseteq \cup_i x_{p_i}^{-1}((-1, 1)^m).$$

With the aid of a partition of unity, as in Lemma 2.1.19, one finds smooth functions $\varphi_i: M \longrightarrow \mathbb{R}$ with $0 \leq \varphi_i \leq 1$ and $\operatorname{supp} \varphi_i \subseteq U_{p_i}$, such that $\sum_i \varphi_i = 1$ on $\cup_i U_{p_i}$. We then set $\omega_i := \varphi_i \cdot \omega$, and thus $\omega = \sum \omega_i$. The derivative and the integral are linear, so it suffices to prove the claim for the ω_i. In other words, we can assume that $\omega = \omega_i$ for some $1 \leq i \leq n$. We now set $V = (-2, 2)^m$, $W = U_{p_i}$ and $h = x_{p_i}^{-1}$. Then $h: V \longrightarrow W$ is a diffeomorphism.

We now discuss the cases $p_i \in D$ and $p_i \in \partial D$ separately. In the first case, we set $Q = [-1, 1]^m$ and $P = f(Q)$. Then $\operatorname{supp} \omega$ is in the interior of P and with the integral

---

[8]Guido Fubini (1879–1943).

formula 3.7.5, we therefore obtain

$$\int_D d\omega = \int_P d\omega = \int_{\partial_R P} \omega = 0 = \int_{\partial D} \omega.$$

For $p_i \in \partial D$ we set $Q = [-1, 0] \times [-1, 1]^{m-1}$ and, again, $P = f(Q)$. The support supp ω only intersects the boundaries of P and D at $h(\{0\} \times (-1, 1)^{m-1})$, and the orientations induced by P and D on these parts of their boundaries agree. With the integration formula 3.7.5, it therefore also follows in this case that

$$\int_D d\omega = \int_P d\omega = \int_{\partial_R P} \omega = \int_{\partial D} \omega.$$ □

In many applications, the boundary of the domain is not smooth, but the singular part of the boundary is not too large, so Stokes's integral formula still holds. We even treated the case of cuboids, to which we reduced the smooth case, first. Many other cases can be reduced to this one with appropriate coverings and decompositions. Further considerations can be found in, for example, section III.14 of [Wh2] or section XVIII.6 of [La].

In Stokes's integral formula 3.7.6, we can choose $D = M$. Then $\partial D = \emptyset$, and therefore the right side of the formula vanishes.

Corollary 3.7.7 *Let M be oriented of dimension m and ω be a smooth $(m-1)$-form on M with compact support. Then $\int_M d\omega = 0$.*   □

Corollary 3.7.8 *Let M be compact, oriented, and have dimension m, and let ω_0 and ω_1 be closed, cohomologous m-forms on M. Then*

$$\int_M \omega_0 = \int_M \omega_1.$$

Proof
With $\omega_1 - \omega_0 = d\eta$, the claim follows from Corollary 3.7.7.   □

Corollary 3.7.9 *Let M be compact, oriented, and have dimension m. Let $f_0, f_1 : M \longrightarrow N$ be homotopic smooth maps and ω be a closed m-form on N. Then*

$$\int_M f_0^* \omega = \int_M f_1^* \omega.$$

Proof
By Corollary 3.4.4, $f_0^* \omega$ and $f_1^* \omega$ are closed, cohomologous m-forms on M, and therefore the claim follows from Corollary 3.7.8.   □

3.8 Supplementary Literature

We did not introduce de Rham cohomology with compact support. It appears in Exercise 3.9.1, and, in light of this, the reader may wish to find their own definition. This and more on differential forms, orientations, and de Rham cohomology can be found in [BT], [Sp1, Chapter 6–8], and [ST].

The classical integral formulas of vector analysis, as used in physics, are special cases of Stokes' integral formula 3.7.6. The role of differential forms in geometry and physics is discussed in [AF].

3.9 Exercises

Exercise 3.9.1
Let $\omega = \varphi dx$ be a Pfaffian form on $\mathbb{R}$, where φ is smooth with compact support. Then $\int_{-n}^{n} \omega = 0$ for all sufficiently large n if and only if ω admits a potential with compact support. ∎

Exercise 3.9.2
On the (x, y)-plane $\mathbb{R}^2$, let $\alpha = xdy - ydx$ and $c: [a, b] \longrightarrow \mathbb{R}^2$ be piecewise smooth, with $c(a) = c(b)$. Then $\int_c \alpha = 2F$, where F is the "oriented surface area" of the region of the plane determined by c, computed "with multiplicity". Compute the integral of α over the boundary curves of rectangles and disks. ∎

Exercise 3.9.3
The winding form ω on $\mathbb{R}^2 \setminus \{0\}$ satisfies Eq. (3.10), but has no primitive function on $\mathbb{R}^2 \setminus \{0\}$. Hint: consider the integral of the winding form along the closed curve $(\cos t, \sin t), 0 \le t \le 2\pi$. ∎

Exercise 3.9.4
Let $W \subseteq M$ be open and ω be a Pfaffian form on W. Then ω is smooth if and only if the function $W \longrightarrow \mathbb{R}, p \mapsto \omega_p(X(p))$, is smooth for all smooth vector fields X on W. ∎

Exercise 3.9.5
Let W be open in M and ω be a k-form on W. Then ω is smooth if and only if the function

$$\omega(X_1, \ldots, X_k): W \longrightarrow \mathbb{R}, \quad p \mapsto \omega_p(X_1(p), \ldots, X_k(p)), \tag{3.41}$$

is smooth for all smooth vector fields $X_1, \ldots, X_k$ on W. ∎

Exercise 3.9.6
Determine the domains on which $f: \mathbb{R}^2 \longrightarrow \mathbb{R}^2$, $f(r, \varphi) = (r \cos \varphi, r \sin \varphi)$, is a diffeomorphism, and compute $f^* dx^1$ and $f^* dx^2$. ∎

Exercise 3.9.7

1. For the *volume form* $\omega = dx^1 \wedge \cdots \wedge dx^m$ on $\mathbb{R}^m$,

$$\omega(v_1, \ldots, v_m) = \det(v_1, \ldots, v_m),$$

where the right hand side means the determinant of the matrix which has the v_i as columns.

2. Compare the $(m-1)$-form

$$\alpha = \sum_{1 \leq i \leq m} (-1)^{i-1} x^i dx^1 \wedge \cdots \wedge \widehat{dx^i} \wedge \cdots \wedge dx^m$$

on $\mathbb{R}^m$ with the form of the same name from Exercise 3.9.2, and show that

$$\alpha_x(v_1, \ldots, v_{m-1}) = \omega(x, v_1, \ldots, v_{m-1})$$

for all $x \in \mathbb{R}^m$ and $v_1, \ldots, v_{m-1} \in T_x \mathbb{R}^m \cong \mathbb{R}^m$. Also show that $d\alpha = m\omega$ and $r^* \alpha = \|x\|^{-m} \alpha$ on $\mathbb{R}^m \setminus \{0\}$, where $r = r(x) = x/\|x\|$. ∎

Exercise 3.9.8
The so-called *symplectic form*

$$\omega = dx^1 \wedge dy^1 + \cdots + dx^n \wedge dy^n$$

on $\mathbb{R}^{2n} = \{(x^1, y^1, \ldots, x^n, y^n) \mid x^1, y^1, \ldots, x^n, y^n \in \mathbb{R}\}$ is the central object in symplectic geometry. Show:

1. $d\omega = 0$.
2. There is a Pfaffian form α with $d\alpha = \omega$ (thereby strengthening 1.).
3. ω is not degenerate: For all $v \in \mathbb{R}^{2n}$, $v \neq 0$, there is a $w \in \mathbb{R}^{2n}$ with $\omega(v, w) \neq 0$.
4. By writing $z^j = x^j + iy^j$, $\mathbb{R}^{2n} \cong \mathbb{C}^n$ (as real vector spaces). The multiplication by i is therefore an isomorphism of $\mathbb{R}^{2n}$, which we denote by J (as is conventional). In this notation, $\langle v, w \rangle := \omega(v, Jw)$, $v, w \in \mathbb{R}^{2n}$, is the Euclidean scalar product (strengthening 3.). Conversely, $\omega(v, w) = \langle Jv, w \rangle$.
5. $\omega^n = \omega \wedge \cdots \wedge \omega$ (n times) $= n! \, dx^1 \wedge dy^1 \wedge \cdots \wedge dx^n \wedge dy^n$. ∎

Exercise 3.9.9 (To Be Worked Before ▶ Sect. 3.4)

1. If M is connected, then $H^0(M) \cong \mathbb{R}$. Also determine $H^0(M)$ in non-connected cases.
2. For all $m \geq 1$, $H^1(\mathbb{R}^m) = \{0\}$.
3. For all $m \geq 2$, $H^1(S^m) = \{0\}$. Hint: with respect to the charts $(U_\pm, \pi_\pm)$ as in Example 2.1.2.2, any closed 1-form ω on S^m has a potential $f_\pm$ on $U_\pm$ by 2. Compare these on $U_+ \cap U_-$.

4. $H^1(S^1) \cong \mathbb{R}$. The torus $T^m = (S^1)^m$ has first Betti number $b_1(T^m) \geq m$. (In fact, it is even true that $b_1(T^m) = m$.) For $m \geq 2$, S^m and T^m are not diffeomorphic.

■

Exercise 3.9.10

1. For $a < b$ and $t \in (a, b)$, $i_t : M \longrightarrow M \times (a, b)$, $i_t(p) := (p, t)$, is a homotopy equivalence.
2. The inclusion $i : S^{m-1} \longrightarrow \mathbb{R}^m \setminus \{0\}$ is a homotopy equivalence.
3. Show that M is contractible if and only if, for all points (or, equivalently, for one point) $p \in M$, the inclusion $i : \{p\} \longrightarrow M$ is a homotopy equivalence. Compare the Poincaré Lemma with Corollary 3.4.5.
4. Let $H(x, t) = tx$ be the contraction of $\mathbb{R}^m$ from Example 3.4.6.1, and $\omega = f \, dx^{i_1} \wedge \cdots \wedge dx^{i_k}$ a k-form on $\mathbb{R}^m$. Then

$$IH^*\omega = \int_0^1 t^{k-1} f(tx) dt \sum_j (-1)^{j-1} x^{i_j} dx^{i_1} \wedge \cdots \wedge \widehat{dx^{i_j}} \wedge \cdots \wedge dx^{i_k}. \qquad (3.42)$$

Compare the sum on the right-hand side with the differential form α from Exercise 3.9.7.2.

■

Exercise 3.9.11
Using an argument like that of Example 3.5.3 show that

$$H^k(\mathbb{K}P^n) \cong \begin{cases} \mathbb{R} & \text{for } k \in \{0, d, 2d, \dots, nd\}, \\ 0 & \text{otherwise} \end{cases}$$

for $\mathbb{K} \in \{\mathbb{C}, \mathbb{H}\}$ and $d = \dim_{\mathbb{R}} \mathbb{K} \in \{2, 4\}$.

■

Exercise 3.9.12
For open subsets W_1, W_2 in M, let $f : W_1 \cap W_2 \longrightarrow W_1 \cup W_2$ be the inclusion. Show: For the de Rham cohomology classes $a \in H^k(W_1 \cap W_2)$ and $b \in H^l(W_1 \cup W_2)$, $\delta(a \wedge f^*b) = \delta a \wedge b$ in $H^{k+l+1}(W_1 \cup W_2)$. Hint: Examine the definition of δ in Appendix B and consider what each choice means in the case of differential forms.

■

Exercise 3.9.13

1. A manifold is orientable if and only if each of its connected components is orientable.
2. If $f : M \longrightarrow N$ is a local diffeomorphism and N is oriented, then M has exactly one orientation such that f is orientation-preserving.
3. The real projective space $\mathbb{R}P^m$ is orientable if and only if m is odd. To this end, consider that $\mathbb{R}P^m$ is orientable if and only if the antipodal map on S^m is orientation-preserving.
4. Lie groups are orientable.

■

Exercise 3.9.14

1. If $f: U \longrightarrow V$ is a diffeomorphism between open subsets of $\mathbb{R}^m$, then f is orientation-preserving if and only if $\det Df > 0$.
2. Let f be a holomorphic function on $\mathbb{C}$, $U = \{y \in \mathbb{C} \mid (df/dz)(z) \neq 0\}$, and $V = f(U)$. Then U and V are open subsets of $\mathbb{R}^2 \cong \mathbb{C}$ and $f: U \longrightarrow V$ is an orientation-preserving local diffeomorphism.

∎

Exercise 3.9.15

1. If $\mathcal{A}$ is an atlas on M such that the change of coordinates $y \circ x^{-1}$ is orientation-preserving for all charts (U, x) and (V, y) in $\mathcal{A}$, then M has an orientation with respect to which the charts in $\mathcal{A}$ are positively oriented.
2. If M is orientable, then M has a locally finite atlas (U_i, x_i) of positively oriented charts together with a partition of unity (φ_i) (with $\operatorname{supp} \varphi_i \subseteq U_i$). If ω_i is the smooth m-form on M with $\omega_i = \varphi_i dx_i^1 \wedge \cdots \wedge dx_i^m$ on U_i and $\omega_i = 0$ outside of $\operatorname{supp} \varphi_i$, then $\omega = \sum \omega_i$ is a smooth m-form on M. For all $p \in M$ and $v_1, \ldots, v_m \in T_p M$, $\omega_p(v_1, \ldots, v_m) > 0$ if and only if $(v_1, \ldots, v_m)$ is a positively oriented basis of $T_p M$.
3. A smooth m-form ω on M such that $\omega_p \neq 0$ for all $p \in M$ is called an *orientation form* or a *volume form*. For $p \in M$, we call a basis $(v_1, \ldots, v_m)$ of $T_p M$ positively oriented, if $\omega_p(v_1, \ldots, v_m) > 0$. This determines an orientation on M in the sense of Definition 3.6.1.
4. If ω is an orientation form on M, then for every smooth m-form η on M there is a smooth function φ on M with $\eta = \varphi \omega$.
5. If $f: M \longrightarrow N$ is a local diffeomorphism between oriented manifolds and ω_M and ω_N are (associated) orientation forms on M and N, then f is orientation-preserving if and only if $f^* \omega_N = \varphi \omega_M$ with $\varphi > 0$.

∎

Exercise 3.9.16

For an embedding $f: M \longrightarrow S^m$ as in the Jordan-Brouwer Separation Theorem 3.6.7, the boundary of the connected components of $S^m \setminus f(M)$ consists of the union of the connected components of $f(M)$. There are at least two and at most z connected components of $S^m \setminus f(M)$, whose boundary consists of exactly one connected component of $f(M)$. ∎

Exercise 3.9.17

1. Compare the integral of differential forms with the integral of Pfaffian forms in ▶ Sect. 3.1.
2. Integrate the $(m-1)$-form α from Exercise 3.9.7.2 over the boundary of compact cuboids. Compare with Exercise 3.9.2.

3. For α and ω as in Exercise 3.9.7, all smooth functions φ on $\mathbb{R}^m$, and all bounded domains $D \subseteq \mathbb{R}^m$ with smooth boundary ∂D,

$$\int_D \varphi \omega = \int_{\partial D} \psi \alpha$$

with $\psi(x) = \int_0^1 t^{m-1} \varphi(tx)\, dt$. Hint: compare with Exercise 3.9.10.4.

4. A smooth m-form ω on S^m is exact if and only if $\int_{S^m} \omega$ vanishes. Hint: See (3.35). ∎

Exercise 3.9.18

For manifolds M and N of dimensions m and n, orientations on M and N induce a canonical orientation on $M \times N$. For any m-form α on M and n-form β on N,

$$\int_{M \times N} \pi_M^* \alpha \wedge \pi_N^* \beta = \int_M \alpha \int_N \beta,$$

where π_M and π_N denote the projections to M and N. ∎

The Geometry of Submanifolds

Werner Ballmann

© Springer Basel 2018
W. Ballmann, *Introduction to Geometry and Topology*, Compact Textbooks in Mathematics,
https://doi.org/10.1007/978-3-0348-0983-2_4

In this chapter we will discuss the geometry of submanifolds of Euclidean spaces. We assume that the reader is familiar with the fundamentals of Euclidean geometry, that is, the geometry of $\mathbb{R}^m$ equipped with the Euclidean scalar product and the associated metric $d(p, q) = \|p - q\|$. A motion of $\mathbb{R}^m$ is a map $B \colon \mathbb{R}^m \longrightarrow \mathbb{R}^m$ of the form $B(x) = Sx + t$ for $S \in O(m)$ and $t \in \mathbb{R}^m$. As a warm-up exercise, show that a map $\mathbb{R}^m \longrightarrow \mathbb{R}^m$ preserves distances if and only if it is a motion.

The geometry of submanifolds consists of two parts: interior and exterior geometry. Interior geometry relates to measurements within the submanifold, exterior geometry to the form of the submanifold relative to the surrounding Euclidean space. We will first briefly discuss these two aspects for curves in Euclidean spaces, for which interior geometry naturally consists only of measuring the lengths of curve segments. Afterwards, we will consider the interior geometry of submanifolds, then their exterior geometry, and conclude by proving the Theorema Egregium of Gauß.[1]

We are interested in *geometric properties and invariants* of submanifolds $M \subseteq \mathbb{R}^n$. These should not depend on the coordinates of M, and should not change if M is replaced by $B(M)$, where B is a motion of $\mathbb{R}^n$. The lengths of curves are examples of such invariants, but there are less obvious, deeper, properties and invariants.

Throughout, we identify the tangent space $T_x\mathbb{R}^n$ with $\mathbb{R}^n$ as in Example 2.2.3.1, and the tangent spaces of submanifolds $M \subseteq \mathbb{R}^n$ with linear subspaces of $\mathbb{R}^n$ as in Proposition 2.3.2.2. Compare with Example 2.2.3.2. More generally, we consider immersions $f \colon M \longrightarrow \mathbb{R}^n$ instead of submanifolds.

With the above identification of the tangent spaces, a *vector field along f* is, in this chapter, a map $X \colon M \longrightarrow \mathbb{R}^n$. We then imagine $X(p)$ as a vector with basepoint $f(p)$. We call such an f *tangential*, if $X(p) \in \operatorname{im} f_{*p}$ for all $p \in M$, that is, when there is a vector field Y on M with $X = df \circ Y$. For a smooth map φ from M to a vector space V, we also commonly write $X\varphi$ instead of $d\varphi \circ X$. Compare with (2.19).

---

[1] Johann Carl Friedrich Gauß (1777–1855).

4.1   Curves

We begin our investigations with curves in Euclidean spaces. If $c\colon I \longrightarrow \mathbb{R}^n$ is a smooth curve, we typically interpret the parameter $t \in I$ as time, and therefore call $\|\dot{c}\|$ the *speed* of c. We call c *regular*, if $\dot{c}(t) \neq 0$ for all $t \in I$. If c is regular, then the *field of directions* of c,

$$e\colon I \longrightarrow \mathbb{R}^3, \quad e(t) := \dot{c}(t)/\|\dot{c}(t)\|, \tag{4.1}$$

is a smooth vector field along c with constant norm 1. We will sometimes also refer to e as the *direction field* of c.

Example 4.1.1
As curves $c\colon \mathbb{R} \longrightarrow \mathbb{R}^2$,
1. $c(t) = (t, t^3)$ is regular,
2. $c(t) = (t^2, t^3)$ is not regular at $t = 0$ and its image has a crease at $(0,0)$,
3. $c(t) = (t^3, t^3)$ is not regular at $t = 0$, although its image, a line, looks regular.

∎

If $c\colon I \longrightarrow \mathbb{R}^n$ is a smooth curve and $\varphi\colon J \longrightarrow I$ is a diffeomorphism, then $\tilde{c} = c \circ \varphi$ is regular if and only if c is regular. We refer to such a φ as a *change of parameters* and $\tilde{c}$ as a *reparameterization* of c. More generally, we call $c \circ \varphi$ a *monotone reparameterization* of c, if $\varphi\colon [\alpha, \beta] \longrightarrow [a, b]$ is monotone, surjective, and smooth.

4.1.1 Length and Energy

Geometry in Euclidean spaces is based on length measurements. Let $c\colon [a, b] \longrightarrow \mathbb{R}^n$ be a smooth curve. Then

$$L(c) := \int_a^b \|\dot{c}(t)\|\, dt \quad \text{and} \quad E(c) = \frac{1}{2} \int_a^b \|\dot{c}(t)\|^2\, dt \tag{4.2}$$

are respectively called the *length* and *energy* of c. It follows from the Cauchy-Schwarz inequality[2] that

$$L^2(c) = \left(\int_a^b 1 \cdot \|\dot{c}\| \right)^2 \leq \left(\int_a^b 1^2 \right) \left(\int_a^b \|\dot{c}\|^2 \right) = 2(b - a)E(c), \tag{4.3}$$

and equality holds if and only if c has constant speed, that is, when the function $\|\dot{c}\|$ is constant. Clearly, the length and energy of curves are invariant under motions: If B is a motion of $\mathbb{R}^n$, then $L(B \circ c) = L(c)$ and $E(B \circ c) = E(c)$. The lengths of curves

---

[2]Hermann Amandus Schwarz (1843–1921).

are, moreover, independent of the parameterization of the curve, see Exercise 4.6.1.1. Length is therefore a geometric invariant of curves in the sense described above.

We say that a smooth curve $c \colon I \longrightarrow \mathbb{R}^n$ is *parameterized by arc length*, if c has constant speed $\|\dot{c}(t)\| = 1$. Then c is regular, and the lengths of segments of c correspond to the lengths of the parameter intervals.

Proposition 4.1.2 *If $c \colon I \longrightarrow \mathbb{R}$ is a regular curve and $t_0 \in I$, there is a change of parameters $\varphi \colon J \longrightarrow I$ with $0 \in J$ such that $\varphi(0) = t_0$ and $c \circ \varphi$ is parameterized by arc length.*

Proof

Define $\psi \colon I \longrightarrow \mathbb{R}$ via $\psi(t) := \int_{t_0}^{t} \|\dot{c}\|$. Since c is regular, ψ is smooth with $\dot{\psi}(t) = \|\dot{c}(t)\| \neq 0$. Therefore ψ is a diffeomorphism on an interval $J \subseteq \mathbb{R}$ with $\psi(t_0) = 0$. The inverse map $\varphi \colon J \longrightarrow I$ is then the desired change of parameters. □

We now discuss a few elementary facts about length and energy. The arguments are chosen so that they may be used again in later extensions of these claims.

Proposition 4.1.3 *For $x, y \in \mathbb{R}^n$ and all smooth curves $c \colon [a, b] \longrightarrow \mathbb{R}^n$ from x to y, $L(c) \geq d(x, y)$. Equality holds if and only if c is a monotone reparameterization of the line $ty + (1 - t)x$, $0 \leq t \leq 1$, from x to y.*

Proof

We assume $x \neq y$ and set $v := (y - x)/\|y - x\|$, the unit vector pointing from x to y. Let $h \colon \mathbb{R}^n \longrightarrow \mathbb{R}$, $h(z) := \langle v, z \rangle$, be the height function associated to v. Then $\operatorname{grad} h = v$, and therefore

$$L(c) = \int_a^b \|\dot{c}\| \geq \int_a^b \langle v, \dot{c} \rangle = \int_a^b \langle \operatorname{grad} h, \dot{c} \rangle$$

$$= h(c(b)) - h(c(a)) = h(y) - h(x) = \|y - x\| = d(x, y).$$

Equality implies that $\langle v, \dot{c} \rangle = \|\dot{c}\|$. □

Corollary 4.1.4 *For $x, y \in \mathbb{R}^n$,*

$$d(x, y) = \min\{L(c) \mid c \text{ is a smooth curve from } x \text{ to } y\},$$

and the minimum is realized precisely by the monotone reparameterizations of the line segment $ty + (1 - t)x$, $0 \leq t \leq 1$, from x to y. □

Energy possesses better analytic properties than length, it is *not* invariant under change of parameters, but rather prefers curves of constant speed:

Proposition 4.1.5 *Let $x, y \in \mathbb{R}^n$ and*

$$c_0 \colon [a, b] \longrightarrow \mathbb{R}^n, \quad c_0(t) = \frac{1}{b - a}\big((t - a)y + (b - t)x\big),$$

be the line segment from x to y. Then for all smooth curves $c\colon [a, b] \longrightarrow \mathbb{R}^n$ from x to y, $E(c) \geq E(c_0)$, and equality implies $c = c_0$.

Proof

We assume $x \neq y$. With terminology as in Proposition 4.1.3 we then obtain

$$2E(c) = \int_a^b \|\dot{c}\|^2 \geq \int_a^b \langle v, \dot{c} \rangle^2$$

$$\geq \frac{1}{b-a} \left(\int_a^b \langle v, \dot{c} \rangle \right)^2$$

$$= \frac{1}{b-a} \left(h(y) - h(x) \right)^2 = 2E(c_0).$$

Equality implies that $\langle v, \dot{c} \rangle$ is constant and $\dot{c} = \langle v, \dot{c} \rangle v$.   □

4.1.2 Curvature

While the derivative of a curve at a point is the best approximation of the curve by a line, we are now interested in approximations of second order. To this end, let $c\colon I \longrightarrow \mathbb{R}^n$ be a smooth curve and t_0 be a point in the interior of I such that $\dot{c}(t_0)$ and $\ddot{c}(t_0)$ are linearly independent.

Proposition 4.1.6 *For $t_1 < t_2 < t_3$ in I close enough to t_0, $c(t_1)$, $c(t_2)$ and $c(t_3)$ are not collinear. Moreover, for $t_1, t_2, t_3 \longrightarrow t_0$, the unique circle $K(t_1, t_2, t_3)$ through $c(t_1)$, $c(t_2)$, and $c(t_3)$ converges to a limit circle $K(t_0)$; see ◘ Fig. 4.1. This limit circle passes through $c(t_0)$, is tangent to c at $c(t_0)$, and lies in the affine plane $E(t_0)$ through $c(t_0)$ spanned by $\dot{c}(t_0)$ and $\ddot{c}(t_0)$. Its center $M = M(t_0) \in E(t_0)$ is determined by the following system of linear equations:*

$$\langle M - c(t_0), \dot{c}(t_0) \rangle = 0,$$

$$\langle M - c(t_0), \ddot{c}(t_0) \rangle = \|\dot{c}(t_0)\|^2.$$

◘ **Fig. 4.1** Circles
approximating a curve

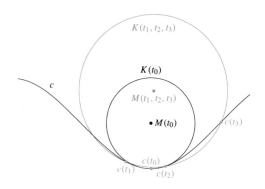

Proof

The first claim follows easily from the second-order Taylor approximation[3] to c at t_0. We now denote by $M(t_1, t_2, t_3)$ the center of the circle $K(t_1, t_2, t_3)$. Then the function

$$f = f(t) := \|c(t) - M(t_1, t_2, t_3)\|^2$$

takes the same value at the points t_1, t_2, and t_3, namely, the square of the radius of $K(t_1, t_2, t_3)$. Its first derivative

$$2\langle c(t) - M(t_1, t_2, t_3), \dot{c}(t) \rangle$$

therefore has zeros at points $s_1 \in (t_1, t_2)$ and $s_2 \in (t_2, t_3)$, and so its second derivative

$$2\langle c(t) - M(t_1, t_2, t_3), \ddot{c}(t) \rangle + 2\|\dot{c}(t)\|^2$$

has a zero at a point $r \in (s_1, s_2)$. □

Definition 4.1.7

For c and t_0 as above, we call $E(t_0)$ the *osculating plane* and $K(t_0)$ the *osculating circle* of c at t_0. We call $M(t_0)$ the *center of curvature*, $R(t_0) = \|M - c(t_0)\|$ the *radius of curvature*, and $\kappa(t_0) = 1/R(t_0)$ the *curvature* of c at t_0.

In light of the first equation in Proposition 4.1.6, the second equation says that $R(t_0)$ times the norm of the component of $\ddot{c}(t_0)$ perpendicular to $\dot{c}(t_0)$ is equal to $\|\dot{c}(t_0)\|^2$. In other words,

$$\kappa(t_0) = \frac{\|\ddot{c}(t_0) - \langle e(t_0), \ddot{c}(t_0) \rangle e(t_0)\|}{\|\dot{c}(t_0)\|^2}, \tag{4.4}$$

where $e = \dot{c}/\|\dot{c}\|$ denotes the field of directions of c as in (4.1). We thereby obtain a second definition of the curvature $\kappa(t_0)$ of c at t_0, consistent with the first, that also applies when $\ddot{c}(t_0)$ is linearly dependent on $\dot{c}(t_0)$.

If c is parameterized by arc length, then $\langle \dot{c}, \dot{c} \rangle/2$ is by definition constant, and therefore has derivative $\langle \dot{c}, \ddot{c} \rangle = 0$. Then $\ddot{c}(t)$ is perpendicular to $\dot{c}(t)$ for all $t \in I$ and consequently

$$\kappa(t_0) = \|\ddot{c}(t_0)\|. \tag{4.5}$$

In Exercises 4.6.2 and 4.6.3 we will check that the curvature of curves is a geometric invariant, which adequately replicates our conception of curvature.

---

[3] Brook Taylor (1685–1731).

4.1.3 Plane Curves

We call curves $c\colon I \longrightarrow \mathbb{R}^2$ *plane curves*. If $c\colon I \longrightarrow \mathbb{R}^2$ is a regular plane curve, then there is exactly one vector field n along c perpendicular to the field of directions $e = \dot{c}/\|\dot{c}\|$ of c such that $(e(t), n(t))$ is a positively oriented basis of $\mathbb{R}^2$ for all $t \in I$, namely

$$n = (-e^2, e^1)$$

with $e = (e^1, e^2)$; see ◻ Fig. 4.2. We call $n = n(t)$ the *principal normal (vector) field* of c. This allows us to give a sign to the curvature of regular plane curves,

$$\kappa_o(t) := \frac{\langle n(t), \ddot{c}(t) \rangle}{\|\dot{c}(t)\|^2}. \tag{4.6}$$

We call κ_o the *oriented curvature* of c. We can rewrite the formula (4.6) as

$$\kappa_o(t) = \frac{\det(\dot{c}(t), \ddot{c}(t))}{\|\dot{c}(t)\|^3}. \tag{4.7}$$

If c is parameterized by arc length, we obtain

$$\kappa_o(t) = \langle n(t), \ddot{c}(t) \rangle = \det(\dot{c}(t), \ddot{c}(t)). \tag{4.8}$$

It is clear that, for the curvature defined in the preceding section, $\kappa = |\kappa_o|$.

◻ **Fig. 4.2** The vector fields e and n

Proposition 4.1.8 *For the field of directions e and the principal normal field n of a regular plane curve $c\colon I \longrightarrow \mathbb{R}^2$,*

$$\dot{e} = \|\dot{c}\| \kappa_o n \quad and \quad \dot{n} = -\|\dot{c}\| \kappa_o e.$$

Proof
Since the functions $\langle e, e \rangle$ and $\langle n, n \rangle$ are constant, their derivatives vanish, so $\langle e, \dot{e} \rangle = \langle n, \dot{n} \rangle = 0$. Therefore $\dot{e}$ is a multiple of n and $\dot{n}$ is a multiple of e. Now $\|\dot{c}\| e = \dot{c}$, and therefore $\|\dot{c}\| \langle n, \dot{e} \rangle = \langle n, \ddot{c} \rangle = \|\dot{c}\|^2 \kappa_o$. This proves the first equation. The function $\langle n, e \rangle$ is also constant, so $\langle \dot{n}, e \rangle = -\langle n, \dot{e} \rangle$. The second equation follows.   □

With the aid of the differential equations from Proposition 4.1.8, we obtain further interpretations of the curvature: Let $c\colon I \longrightarrow \mathbb{R}^2$ be a regular plane curve, and $v \in \mathbb{R}^2$

be a fixed unit vector. Let $\alpha = \alpha(t) \in \mathbb{R}/2\pi\mathbb{Z}$ be the oriented angle from v to $e(t)$, that is $\cos\alpha = \langle v, e \rangle$ and $\sin\alpha = -\langle v, n \rangle$. With Proposition 4.1.8, it follows that

$$\dot{\alpha}\sin\alpha = -\langle v, \dot{e} \rangle = \|\dot{c}\|\kappa_o \sin\alpha,$$

$$\dot{\alpha}\cos\alpha = -\langle v, \dot{n} \rangle = \|\dot{c}\|\kappa_o \cos\alpha.$$

Therefore, we have

$$\dot{\alpha} = \|\dot{c}\|\kappa_o. \tag{4.9}$$

If $I = \mathbb{R}$ and c is periodic with period $\omega > 0$, then $e(\omega) = e(0)$, so $\alpha(\omega) = \alpha(0)$ modulo $2\pi\mathbb{Z}$ and therefore

$$\frac{1}{2\pi}\int_0^\omega \kappa_0(t)\|\dot{c}(t)\|\,dt = k \in \mathbb{Z}. \tag{4.10}$$

The number k is called the *winding number* of c (with respect to the given period ω). The winding number counts the number of times e circles the origin during one period. The *Winding Number Theorem* states that the winding number is ± 1 if c is a Jordan curve, that is, the curve does not repeat points except when forced to by the period. See, for example, section 2.2 in [Kl].

A further interpretation of the curvature refers to the change in the length of c when we pass to parallel curves. To this end, let $I = [a, b]$ and

$$c_s : [a, b] \longrightarrow \mathbb{R}^2, \quad c_s(t) := c(t) + sn(t), \quad s \in \mathbb{R}, \tag{4.11}$$

be the family of *curves parallel to* $c = c_0$. Then by Proposition 4.1.8,

$$\dot{c}_s = \dot{c} - s\|\dot{c}\|\kappa_o e = (1 - s\kappa_o)\dot{c}.$$

Therefore the derivative δL of the length $L(c_s)$ as a function of s at $s = 0$ is given by

$$\delta L = -\int_a^b \kappa_o(t)\|\dot{c}(t)\|\,dt. \tag{4.12}$$

We can therefore interpret the oriented curvature as a measure of the change in length of a plane curve in the normal direction.

For a given speed and a given curvature, the equations from Proposition 4.1.8 are a linear system of ordinary differential equations for the fields e and n. In preparation for the next proposition, we write these equations in matrix form,

$$\dot{F} = FS \tag{4.13}$$

with

$$F := \begin{pmatrix} e^1 & n^1 \\ e^2 & n^2 \end{pmatrix} \quad \text{and} \quad S := \|\dot{c}\| \cdot \begin{pmatrix} 0 & -\kappa_o \\ \kappa_o & 0 \end{pmatrix}. \tag{4.14}$$

Here, the coefficients of F and S are functions of $t \in I$, and differentiating F means differentiating the coefficients of F. The important point in (4.13) is the skew-symmetry of S.

Proposition 4.1.9 *Let $T : I \longrightarrow (0, \infty)$ and $\kappa_o : I \longrightarrow \mathbb{R}$ be smooth functions. Then:*
1. *(Existence) For $t_0 \in I$ and $x_0, e_0 \in \mathbb{R}^2$ with $\|e_0\| = 1$, there is a regular plane curve $c : I \longrightarrow \mathbb{R}^2$ with $c(t_0) = x_0$ and $e(t_0) = e_0$ such that T is the speed of c and κ_o is the oriented curvature of c.*
2. *(Uniqueness) For every two regular plane curves $c_1, c_2 : I \longrightarrow \mathbb{R}^2$ with speed T and oriented curvature κ_o, there is precisely one orientation-preserving motion B of $\mathbb{R}^2$ with $c_2 = B \circ c_1$.*

Proof
Let $F : I \longrightarrow \mathbb{R}^{2 \times 2}$ be a solution of the linear ordinary differential equation (4.13) with $\|\dot{c}\|$ replaced by T and with $F(t_0) \in SO(2)$. By the skew-symmetry of S,

$$\frac{d}{dt}\left(FF^*\right) = \dot{F}F^* + F\dot{F}^* = FSF^* + FS^*F^* = FSF^* - FSF^* = 0.$$

It therefore follows that $F(t)$ is in $O(2)$ for all $t \in I$. However, F is a smooth function of t, and therefore so is $\det F$. Since $\det F(t) = \pm 1$ for all $t \in I$ and $\det F(t_0) = 1$, we must have $\det F(t) = 1$ for all $t \in I$. From this it follows that $F(t)$ is in $SO(2)$ for all $t \in I$. We now choose the particular solution $F : I \longrightarrow \mathbb{R}^{2 \times 2}$ with

$$F(t_0) = \begin{pmatrix} e_0^1 & -e_0^2 \\ e_0^2 & e_0^1 \end{pmatrix} \in SO(2).$$

By what we have just proved, $F(t)$ is in $SO(2)$ for all $t \in I$. Moreover, the first column e and the second column n of F satisfy the differential equations from Proposition 4.1.8 (with $\|\dot{c}\|$ replaced by T). Since $F(t) \in SO(2)$, we then have that $(e(t), n(t))$ is a positively oriented basis of $\mathbb{R}^2$ for all $t \in I$. Therefore,

$$c = c(t) := x_0 + \int_{t_0}^{t} T \cdot e$$

is a smooth plane curve with field of directions e, principal normal field n, speed T, and oriented curvature κ_o. Claim 1 of the proposition follows from this.

Now let $c_1, c_2 : I \longrightarrow \mathbb{R}^2$ be curves with speed T and oriented curvature κ_o. Let $B = Ax + a$ be the unique, orientation-preserving motion of $\mathbb{R}^2$ with

$$A(c_1(t_0)) + a = c_2(t_0), \quad A(e_1(t_0)) = e_2(t_0), \quad A(n_1(t_0)) = n_2(t_0),$$

where e_1, e_2, n_1 and n_2 denote the fields of directions and principal normal fields associated to c_1 and c_2. Then $B \circ c_1$ is a curve parameterized by arc length with direction field $A \circ e_1$, principal normal field $A \circ n_1$, and oriented curvature κ_o. Therefore, $A \circ e_1$ and $A \circ n_1$ are solutions to the equations of Proposition 4.1.8 with the same initial conditions as e_2 and n_2. It therefore follows that $A \circ e_1 = e_2$ and $A \circ n_1 = n_2$. Finally, since $A(c_1(t_0)) + a = c_2(t_0)$, it follows that $c_2 = B \circ c_1$. $\qquad \square$

Remark 4.1.10 The Lie algebra of SO(2) (the tangent space at the identity matrix) consists of the skew-symmetric (2×2)-matrices, compare with Example 2.3.8.3. The matrix $S = S(t)$ in (4.13) is skew-symmetric, so $SO(2) \ni B \mapsto BS(t)$ is a left-invariant vector field on SO(2) for all $t \in I$; see also Exercise 2.7.14.1. The solutions of the corresponding differential equations on SO(2) therefore lie in SO(2). The first part of the above proof shows this with elementary methods.

Corollary 4.1.11 *A regular plane curve moves counter-clockwise around a circle of radius $R > 0$ if and only if its oriented curvature is constant with value $1/R$.*

4.1.4 Space Curves

We now move to the next case, $n = 3$, that of *space curves*, i.e. curves in $\mathbb{R}^3$. Let $c: I \longrightarrow \mathbb{R}^3$ be a smooth space curve, such that $\dot{c}(t)$ and $\ddot{c}(t)$ are linearly independent for all $t \in I$. Then c is regular, and, in addition to the field of directions $e = \dot{c}/\|\dot{c}\|$, we obtain a second vector field along c,

$$n := \frac{\ddot{c} - \langle e, \ddot{c} \rangle e}{\|\ddot{c} - \langle e, \ddot{c} \rangle e\|}, \tag{4.15}$$

the *principal normal field* along c. Together with the *binormal field*

$$b = e \times n, \tag{4.16}$$

where $\times$ denotes the cross product, we obtain the *Frenet frame* $e, n, b: I \longrightarrow \mathbb{R}^3$ of c. The triple $(e(t), n(t), b(t))$ is a positively oriented orthonormal basis of $\mathbb{R}^3$ for all $t \in I$. The corresponding differential equations are called the *Frenet-Serret formulas*[4]:

$$\dot{e} = \|\dot{c}\| \kappa \, n, \quad \dot{n} = -\|\dot{c}\| \kappa \, e + \|\dot{c}\| \tau \, b, \quad \dot{b} = -\|\dot{c}\| \tau \, n. \tag{4.17}$$

Here, $\tau = \tau(t) := \langle \dot{n}(t), b(t) \rangle = -\langle n(t), \dot{b}(t) \rangle$ is called the *torsion* of c.

As in the case of plane curves, it is instructive and helpful to write the Frenet-Serret formulas in matrix form,

$$\dot{F} = FS \tag{4.18}$$

with

$$F := \begin{pmatrix} e^1 & n^1 & b^1 \\ e^2 & n^2 & b^2 \\ e^3 & n^3 & b^3 \end{pmatrix} \quad \text{and} \quad S := \|\dot{c}\| \cdot \begin{pmatrix} 0 & -\kappa & 0 \\ \kappa & 0 & -\tau \\ 0 & \tau & 0 \end{pmatrix}. \tag{4.19}$$

The skew-symmetry of S is again important. Using it, we obtain a theorem about space curves analogous to Proposition 4.1.9:

---

[4] Jean Frédéric Frenet (1816–1900), Joseph Alfred Serret (1819–1885).

Proposition 4.1.12 *Let $T, \kappa: I \longrightarrow (0, \infty)$ and $\tau: I \longrightarrow \mathbb{R}$ be smooth functions.*

1. *(Existence) For $t_0 \in I$ and $x_0, e_0, n_0 \in \mathbb{R}^3$ with (e_0, n_0) orthonormal, there is a regular space curve $c: I \longrightarrow \mathbb{R}^3$ with $c(t_0) = x_0$, $e(t_0) = e_0$ and $n(t_0) = n_0$, such that T is the speed of c, κ the curvature, and τ the torsion.*

2. *(Uniqueness) For every two regular space curves $c_1, c_2: I \longrightarrow \mathbb{R}^3$ with speed T, curvature κ, and torsion τ, there is exactly one orientation-preserving motion B of $\mathbb{R}^3$ with $c_2 = B \circ c_1$.*

Proof

The proof is, *mutatis mutandis*, identical to the proof of Theorem 4.1.9: In some places, the numeral 2 is replaced by 3; the matrices F and S are now (3×3)-matrices as in (4.19), and the initial conditions read

$$F(t_0) := (e_0, n_0, e_0 \times n_0) \in \mathrm{SO}(3).$$

We leave reworking the details as an exercise.   □

4.1.5 Parallelism from a New Perspective

We can only define the Frenet frame along a space curve c if $\dot{c}$ and $\ddot{c}$ are pointwise linearly independent. Moreover, the differential equations (4.17) of Frenet and Serret involve third derivatives of c. An analogous theory in arbitrary dimensions, in particular for curves in $\mathbb{R}^n$ with $n \geq 4$, involves the first n derivatives of the curve and requires that the first $n - 1$ thereof are pointwise linearly independent. Compare, for example, with sections 1.2 and 1.3 in [Kl]. This is rather a lot to ask when n is large.

To arrive at a new canonical class of vector fields along c, we bring another idea into play, that of curve-dependent parallelism. This is a central notion in modern differential geometry, that we first encounter here in a simple situation. To this end, let $c: I \longrightarrow \mathbb{R}^n$ be a regular curve.

Definition 4.1.13

We call a vector field $X: I \longrightarrow \mathbb{R}^n$ along c a *normal field*, if X is perpendicular to c, i.e., if $\langle \dot{c}(t), X(t) \rangle = 0$ for all $t \in I$. For a normal field X along c, we call the component of $\dot{X}$ normal to c the *covariant derivative* of X along c, written as $\nabla X / dt$ or X':

$$\frac{\nabla X}{dt} = X' := \dot{X} - \langle e, \dot{X} \rangle e.$$

We call a normal field X along c *parallel* if $X' = 0$.

If X is a normal field along c, then the function $\langle e, X \rangle$ vanishes identically, and therefore so does its derivative. We thereby obtain

$$\langle e, \dot{X} \rangle = -\langle \dot{e}, X \rangle. \tag{4.20}$$

The component $\langle e(t), \dot{X}(t) \rangle e(t)$ of $\dot{X}(t)$ tangent to c therefore only depends on the value of X at t. It moreover follows that, for given $t \in I$ and $v \in \mathbb{R}^n$ perpendicular to $e(t)$, the derivative of a normal field X along c with $X(t) = v$ has norm at least $|\langle \dot{e}(t), v \rangle|$, with equality holding if and only if $\dot{X}(t)$ is tangent to c. Parallel normal fields are, in this sense, the most parsimonious normal fields.

Proposition 4.1.14 *Let X and Y be normal fields along c. Then the following hold:*
1. *(Linearity) For $\alpha, \beta \in \mathbb{R}$, $(\alpha X + \beta Y)' = \alpha X' + \beta Y'$.*
2. *(Product rule) The function $\langle X, Y \rangle$ has derivative $\langle X', Y \rangle + \langle X, Y' \rangle$.*

Proof
Equation (1) is clear. To prove (2), we compute

$$\frac{d}{dt} \langle X, Y \rangle = \langle \dot{X}, Y \rangle + \langle X, \dot{Y} \rangle = \langle X', Y \rangle + \langle X, Y' \rangle,$$

where we use $\langle e, X \rangle = \langle e, Y \rangle = 0$ for the equality on the right. □

Corollary 4.1.15 *For parallel normal fields X and Y along c, the following hold:*
1. *Linear combinations $\alpha X + \beta Y$ are themselves parallel along c.*
2. *The function $\langle X, Y \rangle$ is constant.* □

Proposition 4.1.16 *For $t_0 \in I$ and $X_0 \in \mathbb{R}^n$ perpendicular to $\dot{c}(t_0)$, there is precisely one parallel normal field X along c with $X(t_0) = X_0$.*

Proof
By Definition 4.1.13 and (4.20), parallel normal fields X along c satisfy the linear ordinary differential equation

$$\dot{X} = -\frac{\langle \ddot{c}, X \rangle}{\|\dot{c}\|^2} \dot{c}. \tag{4.21}$$

Conversely, let $X : I \longrightarrow \mathbb{R}^n$ be the solution of (4.21) with $X(t_0) = X_0$. Scalar multiplication of the left- and right-hand sides of (4.21) by $\dot{c}$ shows that $\langle \dot{c}, X \rangle$ has vanishing derivative, and therefore is constant. Since $\langle \dot{c}(t_0), X(t_0) \rangle = 0$, we conclude from this that X is a normal field along c. In summation, then, it follows that a vector field X along c is a parallel normal field along c if and only if it is a solution of (4.21) and $X(t)$ is perpendicular to $\dot{c}(t)$ for some $t \in I$. □

Corollary 4.1.17 *Let $c: I \longrightarrow \mathbb{R}^n$ be a regular curve. For $t_0 \in I$ and orthonormal vectors $x_2, \ldots, x_n \in \mathbb{R}^n$ perpendicular to $\dot{c}(t_0)$, let $X_2, \ldots, X_n$ be the parallel normal fields along*

c with $X_i(t_0) = x_i$ for all $2 \leq i \leq n$. Then $(e(t), X_2(t), \ldots, X_n(t))$ is an orthonormal basis of $\mathbb{R}^n$ for all $t \in I$. □

A parallel normal field along c does not generally consist of vectors that are parallel in the usual sense—if that were the case, the normal field would be constant. To determine the parallel normal fields along c, one must solve the differential equation (4.21).

Example 4.1.18

Let c be a space curve such that $\dot{c}$ and $\ddot{c}$ are pointwise linearly independent. The Frenet frame $(e, n, b = e \times n)$ is then easy to determine. Determining the parallel normal fields is more difficult. We can express the normal fields as linear combinations $X = \alpha n + \beta b$ with real functions α and β. With the aid of (4.17), the condition that X be parallel then translates to the differential equations

$$\dot{\alpha} = \|\dot{c}\| \tau \beta \quad \text{and} \quad \dot{\beta} = -\|\dot{c}\| \tau \alpha$$

for the coefficients. If the speed and torsion of c are constant, then this system of equations is easy to solve: If α solves the differential equation $\ddot{\alpha} + \|\dot{c}\| \tau \alpha = 0$, then the pair α and $\beta := \dot{\alpha}/\|\dot{c}\|\tau$ solves the system of differential equations above. ∎

Proposition 4.1.19 *For smooth functions* $T : I \longrightarrow (0, \infty)$ *and* $\kappa_2, \ldots, \kappa_n : I \longrightarrow \mathbb{R}$, *the following hold:*

1. *(Existence) For* $t_0 \in I$ *and* $x_0, \ldots, x_n \in \mathbb{R}^n$ *with* $(x_1, \ldots, x_n)$ *orthonormal, there exist a regular curve* $c : I \longrightarrow \mathbb{R}^n$ *with speed* T *and parallel normal fields* $X_2, \ldots, X_n$ *along* c *with* $c(t_0) = x_0$, $e(t_0) = x_1$, $X_i(t_0) = x_i$, *and* $\langle \dot{e}, X_i \rangle = \kappa_i T$ *for all* $2 \leq i \leq n$.
2. *(Uniqueness) For every two regular curves* $c_1, c_2 : I \longrightarrow \mathbb{R}^n$ *with speed* T *and parallel and pointwise orthonormal normal fields* $X_2, \ldots, X_n$ *along* c_1 *and* $Y_2, \ldots, Y_n$ *along* c_2 *with* $\langle \dot{e}_1, X_i \rangle = \kappa_i T$ *and* $\langle \dot{e}_2, Y_i \rangle = \kappa_i T$, *there is exactly one motion* $B = Ax + a$ *of* $\mathbb{R}^n$ *with* $c_2 = B \circ c_1$ *and* $Y_i = A \circ X_i$ *for all* $2 \leq i \leq n$.

Proof

We once again write the differential equations for e and $X_2, \ldots, X_n$ in matrix form,

$$\dot{F} = FS \tag{4.22}$$

with

$$F := \begin{pmatrix} e^1 & X_2^1 & \cdots & X_n^1 \\ \vdots & \vdots & & \vdots \\ e^n & X_2^n & \cdots & X_n^n \end{pmatrix} \quad \text{and} \quad S := T \cdot \begin{pmatrix} 0 & -\kappa_2 & \cdots & -\kappa_n \\ \kappa_2 & 0 & \cdots & 0 \\ \vdots & \vdots & & \vdots \\ \kappa_n & 0 & \cdots & 0 \end{pmatrix}. \tag{4.23}$$

The skew-symmetry of S is again important. The rest proceeds as before. □

Compare the matrices S from (4.19) and (4.23) and the corresponding systems of differential equations (4.18) and (4.22). This "small change" to the matrix S has a large effect, namely that we now have canonical normal fields along regular curves in Euclidean spaces regardless of the dimension of the space. Moreover, the differential equation (4.23) only involves second derivatives of c (and first derivatives of the X_i). We will end with this insight, and not delve any deeper, as we only introduced the concepts of covariant derivatives and parallelism along curves to give motivation for analogous concepts in our later discussions.

4.2 Interior Geometry

After our introductory remarks about curves, we now consider the general case of submanifolds in the Euclidean space $\mathbb{R}^n$. There are two ways these may be specified: firstly via equations, that is as level sets of mappings, or via embeddings or immersions $M \longrightarrow \mathbb{R}^n$. The second case includes the first, since we can think of a submanifold $M \subseteq \mathbb{R}^n$ with inclusion $i : M \longrightarrow \mathbb{R}^n$ as an embedding. Nevertheless, it is often worthwhile to give results separately for the first case, as their formulations are usually simpler and easier to grasp.

The interior geometry of submanifolds deals with measurements of geometric objects which are contained in the submanifold. The measuring stick is provided by the ambient space $\mathbb{R}^n$, namely, the Euclidean scalar product. In every other respect, however, the surrounding space is ignored, in some sense because the inhabitants of the submanifold are incapable of perceiving the space outside the submanifold.[5] In the preceding discussion of curves, their lengths are measurements belonging to their interior geometry, but their curvature is not.

In the following, we denote by M a manifold of dimension m, and by $f : M \longrightarrow \mathbb{R}^n$ an immersion. For $p \in M$, we call the linear subspaces

$$T_p f := \operatorname{im} df(p) \quad \text{and} \quad N_p f := [\operatorname{im} df(p)]^{\perp} \tag{4.24}$$

of $\mathbb{R}^n$ the *tangent space* and *normal space* to f at p. We denote by

$$\pi_p^T : \mathbb{R}^n \longrightarrow \mathbb{R}^n \quad \text{and} \quad \pi_p^N : \mathbb{R}^n \longrightarrow \mathbb{R}^n \tag{4.25}$$

the orthogonal projections from $\mathbb{R}^n$ to $T_p f$ and $N_p f$. If $M \subseteq \mathbb{R}^n$ is a submanifold with inclusion f, then $T_p f = T_p M$ under the usual identification of $T_p M$ as in Proposition 2.3.2. Therefore, we then write $T_p M$ and $N_p M$ instead of $T_p f$ and $N_p f$.

Example 4.2.1
1) For a regular curve $c : I \longrightarrow \mathbb{R}^n$, $T_t c = \mathbb{R} \cdot \dot{c}(t) = \mathbb{R} \cdot e(t)$. For regular plane curves c, $N_t c = \mathbb{R} \cdot n(t)$, for regular space curves c, $N_t c$ is the span of the principal normal $n(t)$ and the binormal $b(t)$.

---

[5]This perceptual difficulty is one of the topics of the novella *Flatland*, written by Edwin Abbott Abbott (1838–1926) and published in 1884.

2) The sphere $S_r^m = \{x \in \mathbb{R}^{m+1} \mid \|x\|^2 = r^2\}$ of radius $r > 0$ is a submanifold of $\mathbb{R}^{m+1}$ with $T_x S^m = \{y \in \mathbb{R}^{m+1} \mid \langle x, y \rangle = 0\}$ and $N_x M = \mathbb{R} \cdot x$ for all $x \in S_r^m$.

∎

Let $X \colon M \longrightarrow \mathbb{R}^n$ be a vector field along f. Then for $p \in M$, we call

$$X^T(p) := \pi_p^T(X(p)) \quad \text{and} \quad X^N(p) = \pi_p^N(X(p)) \tag{4.26}$$

the *tangential* and *normal components* of X at p. We also write

$$X^T = \pi^T \circ X \quad \text{und} \quad X^N = \pi^N \circ X. \tag{4.27}$$

4.2.1 The First Fundamental Form

Measurements in M only depend on the restriction of the Euclidean scalar product of the ambient space $\mathbb{R}^n$ to the tangent spaces $T_p f$. The first fundamental form gets to the heart of this fact.

Definition 4.2.2

For p in M, we call the scalar product g_p on $T_p M$,

$$g_p(v, w) := \langle df(p)(v), df(p)(w) \rangle, \quad v, w \in T_p M,$$

the *first fundamental form* of f at p. We call the family of scalar products $g = (g_p)_{p \in M}$ the *first fundamental form* of f.

Instead of $g_p(v, w)$, we sometimes write $\langle v, w \rangle_p$ or $\langle v, w \rangle$, and similarly for the corresponding norms. These notations are particularly useful for submanifolds $M \subseteq \mathbb{R}^n$, as f is then the inclusion and $df(p)$ is the usual identification of $T_p M$ with a linear subspace of $\mathbb{R}^n$.

We next clarify the regularity of the first fundamental form's dependence on the point p. To this end, let (U, x) be a chart on M. For all $p \in U$, the coordinate fields $\partial/\partial x^1(p), \ldots, \partial/\partial x^m(p)$ then form a basis $T_p M$. Since $df(p)(\partial/\partial x^i(p)) = (\partial f/\partial x^i)(p)$, the coefficients of the fundamental matrix of the first fundamental form with respect to this basis are given by

$$g_{ij}(p) := \Big\langle \frac{\partial}{\partial x^i}(p), \frac{\partial}{\partial x^j}(p) \Big\rangle_p = \Big\langle \frac{\partial f}{\partial x^i}(p), \frac{\partial f}{\partial x^j}(p) \Big\rangle. \tag{4.28}$$

Now the partial derivatives $\partial f/\partial x^i \colon U \longrightarrow \mathbb{R}^n$ are smooth, and therefore so are the functions $g_{ij} \colon U \longrightarrow \mathbb{R}$. Since the g_p are scalar products, the matrix of the g_{ij} is pointwise symmetric and positive definite.

Example 4.2.3

1) If $c : I \longrightarrow \mathbb{R}^n$ is a regular curve, then $g_{tt} = \|\dot{c}\|^2$.[6]

2) *Surfaces of revolution:* Let $c = (r, h) = (r(t), h(t))$, $t \in I$, be a regular curve in the (x, z)-plane with $r > 0$. We call c the *profile curve* of the *surface of revolution*

$$f : I \times \mathbb{R} \longrightarrow \mathbb{R}^3, \quad f(t, \varphi) := (r(t) \cos(\varphi), r(t) \sin(\varphi), h(t)).$$

A concrete example from this class is the torus considered in Exercise 2.7.8.1.

Based on the corresponding terms from geography, we call the curves $\varphi = \text{const}$ *meridians* or *longitudes*, and the curves $t = \text{const}$ *latitudes* of f. The partial derivatives of f are

$$\frac{\partial f}{\partial t} = (\dot{r} \cos(\varphi), \dot{r} \sin(\varphi), \dot{h}), \quad \frac{\partial f}{\partial \varphi} = (-r \sin(\varphi), r \cos(\varphi), 0).$$

Now $\partial f / \partial t$ and $\partial f / \partial \varphi$ are pointwise $\neq 0$ and perpendicular to one another, and are consequently pointwise linearly independent. Therefore, f is an immersion. For all $p \in M$, the vectors $(\partial f / \partial t)(p)$ and $(\partial f / \partial \varphi)(p)$ form a basis of $T_p f$, and their cross product is the generator of the line $N_p f$. The coefficients of the fundamental matrix of the first fundamental form are

$$g_{tt} = \left\langle \frac{\partial f}{\partial t}, \frac{\partial f}{\partial t} \right\rangle = \|\dot{c}\|^2, \quad g_{t\varphi} = g_{\varphi t} = \left\langle \frac{\partial f}{\partial t}, \frac{\partial f}{\partial \varphi} \right\rangle = 0, \quad g_{\varphi\varphi} = \left\langle \frac{\partial f}{\partial \varphi}, \frac{\partial f}{\partial \varphi} \right\rangle = r^2.$$

Therefore, the fundamental matrix

$$\begin{pmatrix} \|\dot{c}\|^2 & 0 \\ 0 & r^2 \end{pmatrix}$$

of the first fundamental form is diagonal. If c is parameterized by arc length, that is, has speed 1, then $g_{tt} = \|\dot{c}\|^2 = 1$.

3) *Generalized helicoids:* Let $a \neq 0$ be a constant and $c = c(t) = (x(t), y(t))$, $t \in I$, be a regular curve in the (x, y)-plane. We call c the *profile curve* of the *generalized helicoid*

$$f(t, \varphi) = (x(t) \cos \varphi - y(t) \sin \varphi, x(t) \sin \varphi + y(t) \cos \varphi, a\varphi).$$

A concrete example in this class is the *helicoid*, where $f(t, \varphi) = (t \cos \varphi, t \sin \varphi, a\varphi)$. The partial derivatives of f are

$$\frac{\partial f}{\partial t} = (\dot{x} \cos \varphi - \dot{y} \sin \varphi, \dot{x} \sin \varphi + \dot{y} \cos \varphi, 0),$$

$$\frac{\partial f}{\partial \varphi} = (-x \sin \varphi - y \cos \varphi, x \cos \varphi - y \sin \varphi, a).$$

---

[6]Here and in other examples we use the variable names as indices.

These are pointwise linearly independent, since $\partial f/\partial t \neq 0$ and $a \neq 0$. The coefficients of the fundamental matrix of the first fundamental form are therefore

$$g_{tt} = \|\dot{c}\|^2, \quad g_{t\varphi} = g_{\varphi t} = x\dot{y} - \dot{x}y, \quad g_{\varphi\varphi} = \|c\|^2 + a^2.$$

These do not depend on the parameter φ.

4) *Ruled surfaces:* Let $c : I \longrightarrow \mathbb{R}^3$ be a regular space curve and $X : I \longrightarrow \mathbb{R}^3$ be a smooth vector field along c, such that $\dot{c}(t)$ and $X(t)$ are linearly independent for all $t \in I$. The associated *ruled surface* is then

$$f(s, t) = c(t) + s X(t).$$

A concrete example is the helicoid as above with $f(s, t) = (s \cos t, s \sin t, at)$.
Each line $t = $ const is called a *generator*, and each curve $s = $ const a *directrix* of f. The partial derivatives of f are

$$\frac{\partial f}{\partial s} = X \quad \text{and} \quad \frac{\partial f}{\partial t} = \dot{c} + s\dot{X}.$$

Therefore, there is a neighborhood U of $\{s = 0\}$ in $\mathbb{R} \times I$, such that f is an immersion on U. The coefficients of the first fundamental form are

$$g_{ss} = \|X\|^2, \quad g_{st} = g_{ts} = \langle X, \dot{c} + s\dot{X}\rangle, \quad g_{tt} = \langle \dot{c} + s\dot{X}, \dot{c} + s\dot{X}\rangle.$$

5) *Graphs:* Let $W \subset \mathbb{R}^m$ be open and $h : W \longrightarrow \mathbb{R}$ be smooth. Then $f : W \longrightarrow \mathbb{R}^{m+1}$, $f(x) := (x, h(x))$, is an embedding with $\partial f/\partial x^j = (e_j, \partial h/\partial x^j)$, where e_j denotes the j-th standard basis vector of $\mathbb{R}^m$. For each $p \in M$, the tangent space $T_p f$ is the span of the $(\partial f/\partial x^j)(p)$. The coefficients of the first fundamental form with respect to this basis are

$$g_{ij} = \langle \frac{\partial f}{\partial x^i}, \frac{\partial f}{\partial x^j}\rangle = \delta_{ij} + \frac{\partial h}{\partial x^i}\frac{\partial h}{\partial x^j}.$$

In the case $m = 2$ the normal spaces to f are generated by the $(\partial f/\partial x^1) \times (\partial f/\partial x^2)$. ∎

Now let (U, x) again be a chart on M, and let X and Y be vector fields on U. We write X and Y as linear combinations

$$X = \xi^i \frac{\partial}{\partial x^i} \quad \text{and} \quad Y = \eta^i \frac{\partial}{\partial x^i}$$

of the coordinate vector fields. At every $p \in U$, then

$$\langle X(p), Y(p)\rangle = \langle \xi^i(p)\frac{\partial}{\partial x^i}(p), \eta^j(p)\frac{\partial}{\partial x^j}(p)\rangle = g_{ij}(p)\xi^i(p)\eta^j(p). \qquad (4.29)$$

For ease of notation, we omit the p, and write (4.29) as an equality of functions,

$$\langle X, Y \rangle = g_{ij}\xi^i\eta^j. \tag{4.30}$$

Since the g_{ij} are smooth, we conclude that the function $\langle X, Y \rangle$ is smooth if X and Y are smooth. We also write

$$g = g_{ij}dx^i dx^j, \tag{4.31}$$

since $dx^i(X) = \xi^i$ and

$$dx^i dx^j(X, Y) := dx^i(X)dx^j(Y) = \xi^i\eta^j. \tag{4.32}$$

We will often need the inverse of the fundamental matrix (g_{ij}). In light of the Einstein summation convention, it is convenient to denote its coefficients by g^{ij}. Then, by definition,

$$g^{ij}g_{jk} = g_{kj}g^{ji} = \delta_k^i. \tag{4.33}$$

The formulas for the inverse matrix from linear algebra show that the g^{ij} are also smooth.

Lemma 4.2.4 *Let $p \in M$ and (U, x) be a chart on M around p. Let $w \in \mathbb{R}^n$ be a vector and $w = w^T + w^N$ be the decomposition of w into tangential and normal components with respect to the decomposition $\mathbb{R}^n = T_p f \oplus N_p f$ as in (4.24). Then there is exactly one tangent vector $v \in T_p M$ with $w^T = df(p)(v)$:*

$$v = \xi^j \frac{\partial}{\partial x^j}(p) \quad \text{and} \quad w^T = \xi^j \frac{\partial f}{\partial x^j}(p) \quad \text{with} \quad \xi^j = \langle w, \frac{\partial f}{\partial x^i}(p)\rangle g^{ij}(p).$$

In particular, the tangential and normal components of a smooth vector field along f are smooth.

Proof
The first claim is clear, since $df(p)$ is injective. Now write $w^T = \xi^j(\partial f/\partial x^j)(p)$. Then

$$\langle w, \frac{\partial f}{\partial x^i}(p)\rangle = \langle w^T, \frac{\partial f}{\partial x^i}(p)\rangle = \xi^k \langle \frac{\partial f}{\partial x^k}(p), \frac{\partial f}{\partial x^i}(p)\rangle = \xi^k g_{ki}(p).$$

From this it follows that

$$\xi^j = \xi^k g_{ki}(p)g^{ij}(p) = \langle w, \frac{\partial f}{\partial x^i}(p)\rangle g^{ij}(p). \qquad \square$$

Remark 4.2.5 In the case of a submanifold $M \subseteq \mathbb{R}^n$, the correspondence between the tangential component w^T of w and tangent vector v of M can be seen as the usual identification of the tangent spaces of M with linear subspaces of $\mathbb{R}^n$, and is, in this sense, trivial.

4.2.2 The Internal Distance

The first fundamental form is defined so that, for all $p \in M$, $df(p) \colon T_pM \longrightarrow \mathbb{R}^n$ preserves norms. We define the length and energy of a piecewise smooth curve $c \colon [a, b] \longrightarrow M$ as before,

$$L(c) = \int_a^b \|\dot{c}(t)\| \, dt \quad \text{and} \quad E(c) = \frac{1}{2} \int_a^b \|\dot{c}(t)\|^2 dt. \tag{4.34}$$

Since $d(f \circ c)/dt = df \circ \dot{c}$, we then have $L(c) = L(f \circ c)$ and $E(c) = E(f \circ c)$.

In Corollary 4.1.4 we saw that the distance between points in Euclidean spaces is realized by the lengths of curves. By analogy, we arrive at the *internal metric* or *internal distance d* on M,

$$d(p, q) = \inf L(c), \quad p, q \in M, \tag{4.35}$$

where the infimum is taken over all piecewise smooth curves c in M running from p to q. The external metric $\|f(p) - f(q)\|$, $p, q \in M$, therefore satisfies $\|f(p) - f(q)\| \leq d(p, q)$ (and is only a true metric if f is injective). The external metric usually requires the use of measurements outside of M (resp. the image of f), as the shortest curves between points in $\mathbb{R}^n$ are lines, which are not typically contained in M (resp. the image of f). See ▢ Fig. 4.3.

▢ **Fig. 4.3** The internal distance and the Euclidean distance between points on the unit sphere

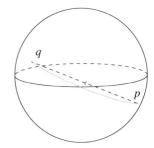

Remark 4.2.6 We allow d to take the value ∞: $d(p, q) = \infty$ if and only if p and q lie in different connected components of M.

Proposition 4.2.7 *The internal metric on M is a metric on M, and it induces the topology of M.*

Proof

Let $p \in M$ and $x \colon U \longrightarrow U'$ be a chart on M around p with $x(p) = 0$. Choose $\varepsilon > 0$ such that the open Euclidean ball B' of radius ε around 0 is contained in U' and such that, for all $q \in x^{-1}(B')$ and $\xi \in \mathbb{R}^m$,

$$\varepsilon^2 \delta_{ij} \xi^i \xi^j \leq g_{ij}(q) \xi^i \xi^j \leq \varepsilon^{-2} \delta_{ij} \xi^i \xi^j.$$

Now, M is a Hausdorff space. Therefore, every continuous path from p to a point q outside of $B = x^{-1}(B')$ must first run through B to $x^{-1}(\partial B')$. Such a path segment has length $\geq \varepsilon^2$. It therefore follows $d(p, q) > 0$.

For the length of a piecewise smooth curve c in B,

$$\varepsilon L(c) \leq L_E(x \circ c) \leq L(c)/\varepsilon,$$

where $L_E(x \circ c)$ denotes the Euclidean distance of $x \circ c$. This estimate carries over to the distance from p to $q \in B$,

$$\varepsilon d(p, q) \leq \|x(p) - x(q)\| \leq d(p, q)/\varepsilon.$$

It therefore follows that $d(p, q) > 0$ whenever $p \neq q$. Since d is symmetric and satisfies the triangle inequality, it is a metric on M. The claim about the topology follows directly from our estimate of distances in B. □

Remark 4.2.8 Since, in general, f need not be injective, we advise caution in the above proof of the positivity of $d(p, q) > 0$.

Example 4.2.9
We use the argument from the proof of Proposition 4.1.3 to show that the internal distance $d(x, y)$ on the sphere S^m of radius 1 in $\mathbb{R}^{m+1}$ is given by the angle $\angle(x, y)$.

For a given $x \in S^m$, we choose $\varphi = \varphi(y) = \angle(x, y)$ as a replacement for the height function h in the proof of Proposition 4.1.3. Now we can write points $y \in S^m \setminus \{\pm x\}$ uniquely as

$$y = \cos(\varphi)x + \sin(\varphi)z$$

with $z = z(y)$ in the equator of x, that is $\langle x, z \rangle = 0$, $\langle z, z \rangle = 1$, and $\varphi = \varphi(y) \in (0, \pi)$. The function $\varphi \colon S^m \setminus \{\pm x\} \longrightarrow (0, \pi)$ is smooth, with gradient (see Exercise 4.6.11)

$$(\operatorname{grad} \varphi)(\cos(\varphi)x + \sin(\varphi)z) = -\sin(\varphi)x + \cos(\varphi)z.$$

We now show that $d(x, y) \geq \varphi(y)$. It suffices to show that every piecewise smooth curve $c \colon [a, b] \longrightarrow S^m$ with $c(a) = x$ has length $L(c) \geq \varphi(c(b))$. To this end, we can assume that $a = \sup\{t \in [a, b] \mid c(t) = x\}$ and $b = \inf\{t \in [a, b] \mid \varphi(c(t)) = \varphi(c(b))\}$. For all $t \in (a, b)$, then, $c(t)$ lies in $S^m \setminus \{\pm x\}$. Since $\|\operatorname{grad} \varphi\| = 1$, we obtain

$$L(c) = \int_a^b \|\dot{c}(t)\| \, dt \geq \int_a^b \langle \operatorname{grad} \varphi(c(t)), \dot{c}(t) \rangle \, dt = \varphi(y).$$

On the other hand, the great circle

$$c \colon [0, 1] \longrightarrow S^m, \quad c(t) = \cos(t\varphi(y))x + \sin(t\varphi(y))z(y)$$

is smooth, with length $\varphi(y) = \angle(x, y)$. The claim that $d(x, y) = \angle(x, y)$ follows. It also follows that the shortest curves connecting two points are monotone reparameterizations of arcs along great circles. ■

4.2.3 Variations and Geodesics

We now address the question of which conditions a piecewise smooth curve $c\colon [a, b] \longrightarrow M$ must satisfy if it is the shortest curve between its endpoints $p = c(a)$ and $q = c(b)$. We consider the length L as a functional on the space of piecewise smooth curves from p to q.[7] The shortest curves are then those for which L attains a minimum, and so will be critical points of L. To make this notion of critical points of L precise, one considers families of piecewise smooth curves, or so-called *variations*, such that L is differentiable along these families. We have already encountered variations in a special case. Compare (4.11) and (4.12).

Definition 4.2.10

A *variation* of c is a map

$$h\colon (-\varepsilon, \varepsilon) \times [a, b] \longrightarrow M \quad \text{with} \quad h(0, t) = c(t),$$

such that there is a subdivision $a = t_0 < t_1 < \cdots < t_k = b$ of $[a, b]$ such that the restriction of h to $(-\varepsilon, \varepsilon) \times [t_{i-1}, t_i]$ is smooth for all $1 \leq i \leq k$.

The variation consists of the piecewise smooth curves $c_s := h(s, .)$. We call the piecewise smooth vector field $V = V(t) = (\partial_s h)(0, t)$ along c the *variation field* of h. Instead of h, we sometimes write (c_s). We call a variation $h = (c_s)$ of $c = c_0$ *proper*, if $c_s(a) = c(a)$ and $c_s(b) = c(b)$ for all $s \in (-\varepsilon, \varepsilon)$. See ☐ Fig. 4.4.

☐ **Fig. 4.4**   A proper variation

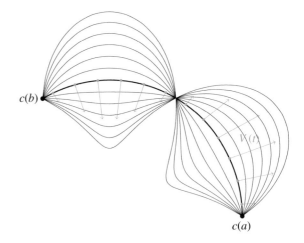

$c(b)$

$V(t)$

$c(a)$

---

[7]Since the domain of L is a space of functions, one does not simply call L a function, but rather uses the hifalutin term *functional*, which does not, however, have a deeper meaning.

Let $c\colon [a, b] \longrightarrow M$ be a piecewise smooth curve. As in the smooth case, we call c *regular* if $\dot{c}(t) \neq 0$ for all $t \in [a, b]$. Now let c be regular and let $h = (c_s)$ be a variation of $c = c_0$. Then the c_s are also regular if we only consider sufficiently small s. By definition of the first fundamental form,

$$\|\dot{c}_s(t)\| = \|\partial_t (f \circ h)(s, t)\|.$$

With a subdivision as in Definition 4.2.10 it therefore follows that $\|\dot{c}_s(t)\|$ is smooth on the rectangles $(-\varepsilon, \varepsilon) \times [t_{i-1}, t_i]$, $1 \leq i \leq k$. From this it follows that the lengths $L(c_s)$ depend smoothly on s.

Definition 4.2.11

Let $c\colon [a, b] \longrightarrow M$ be regular and let $h = (c_s)$ be a variation of $c = c_0$. Then we call the derivative δL of $L(c_s)$ by s at $s = 0$ the *the first variation of arc length* of (c_s).

Proposition 4.2.12 *Let $c\colon [a, b] \longrightarrow M$ be piecewise smooth with constant speed $T \neq 0$. Then for a variation $h = (c_s)$ of $c = c_0$ with variation field V,*

$$\delta L = \frac{1}{T} \left[\sum_{i=1}^{k} \langle V, \dot{c} \rangle \big|_{t_{i-1}}^{t_i} - \int_a^b \left\langle df \circ V, \frac{d^2(f \circ c)}{dt^2} \right\rangle dt \right]$$

$$= \frac{1}{T} \left[\langle V, \dot{c} \rangle \big|_a^b + \sum_{i=1}^{k-1} \langle V(t_i), \Delta_i \rangle - \int_a^b \left\langle df \circ V, \frac{d^2(f \circ c)}{dt^2} \right\rangle dt \right]$$

with $\Delta_i := \dot{c}(t_i-) - \dot{c}(t_i+)$, $1 \leq i \leq k - 1$.

Proof
We may assume that $\dot{c}_s(t) \neq 0$ for all s and t. Then $\|\dot{c}_s(t)\|$ is smooth on the rectangles $(-\varepsilon, \varepsilon) \times [t_{i-1}, t_i]$ as in Definition 4.2.10. Therefore,

$$\frac{d(L(c_s))}{ds} = \frac{d}{ds} \left(\int_a^b \|\dot{c}_s(t)\| \, dt \right)$$

$$= \int_a^b \frac{d}{ds} \sqrt{\langle \partial_t(f \circ h), \partial_t(f \circ h) \rangle} \, dt.$$

With the requirement that $\|\dot{c}(t)\| = T = \text{const}$, we thereby obtain

$$\delta L = \frac{1}{T} \int_a^b \langle \partial_s \partial_t (f \circ h), \partial_t(f \circ h) \rangle (0, t) \, dt$$

$$= \frac{1}{T} \int_a^b \langle \partial_t \partial_s (f \circ h), \partial_t(f \circ h) \rangle (0, t) \, dt$$

$$= \frac{1}{T} \int_a^b \partial_t \langle \partial_s (f \circ h), \partial_t (f \circ h) \rangle (0, t) \, dt - \frac{1}{T} \int_a^b \langle \partial_s (f \circ h), \partial_t^2 (f \circ h) \rangle (0, t) \, dt$$

$$= \frac{1}{T} \sum_{i=1}^k \langle \partial_s (f \circ h), \partial_t (f \circ h) \rangle (0, t) \big|_{t_{i-1}}^{t_i} - \frac{1}{T} \int_a^b \langle \partial_s (f \circ h), \partial_t^2 (f \circ h) \rangle (0, t) \, dt.$$

However, at $s = 0$, by definition,

$$\partial_s (f \circ h) = df \circ V, \quad \partial_t (f \circ h) = df \circ \dot{c} \quad \text{and} \quad \partial_t^2 (f \circ h) = \frac{d^2 (f \circ c)}{dt^2}. \qquad \square$$

We now come to a result of Johann Bernoulli[8] from 1698 (unpublished), one of the first results ever in differential geometry; compare with [HT, pp. 117].

Proposition 4.2.13 *A piecewise smooth curve $c \colon [a, b] \longrightarrow M$ has constant speed T and first variation of arc length $\delta L = 0$ for every proper variation of c if and only if c is smooth and the tangential component of the second derivative of the second derivative of $f \circ c$ vanishes.*

Proof

We first assume that c is smooth and the tangential component of the second derivative of $f \circ c$ vanishes. By the definition of the first fundamental form

$$\frac{d}{dt} \langle \dot{c}, \dot{c} \rangle = \frac{d}{dt} \langle \frac{d(f \circ c)}{dt}, \frac{d(f \circ c)}{dt} \rangle = 2 \langle \frac{d(f \circ c)}{dt}, \frac{d^2 (f \circ c)}{dt^2} \rangle = 0,$$

since $d(f \circ c)/dt$ is tangent to f. Therefore c has constant speed T.

We may now further assume that c is not constant, that is that $T > 0$. Let $h = (c_s)$ be a proper variation of $c = c_0$ with variation field V.

Since h is proper and c is smooth, it follows that $V(a) = 0$, $V(b) = 0$, and $\Delta_i = 0$, $1 \leq i \leq k - 1$. Therefore, the first terms in the formula for δL vanish. The integrand of the integral now vanishes pointwise, since $df \circ V$ is tangential and $d^2 (f \circ c)/dt^2$ is normal to f. Therefore $\delta L = 0$.

In the opposite direction, we can assume that c is not constant. Then let $a = t_0 < t_1 < \cdots < t_k = b$ be a subdivision, such that c is smooth on the intervals $[t_{i-1}, t_i]$. We first show, that the tangential component of the second derivative of $f \circ c$ vanishes on the intervals $[t_{i-1}, t_i]$. Since, by Lemma 4.2.4, the tangential component is smooth on these intervals, it suffices to show that it vanishes on the open intervals (t_{i-1}, t_i). Assume that this is not the case for some $t' \in (t_{i-1}, t_i)$. Then there is a vector $v \in T_{c(t')}M$ with

$$\langle df(c(t'))(v), \frac{d^2 (f \circ c)}{dt^2} (t') \rangle > 0.$$

---

[8]Johann Bernoulli (1667–1748). He, of course, only discussed curves on surfaces in $\mathbb{R}^3$, but this amounts to much the same thing in proofs.

Now let (U, x) be a chart on M around $c(t')$, and let $\xi \in \mathbb{R}^m$ be the principal part of v with respect to x. Then

$$v = \xi^i \frac{\partial}{\partial x^i}\Big|_{c(t')} \quad \text{and} \quad df(c(t')) \cdot v = \xi^i \frac{\partial f}{\partial x^i}(c(t')).$$

Since the derivatives of f and c are continuous, there is an $\varepsilon > 0$ with $(t' - \varepsilon, t' + \varepsilon) \subseteq (t_{i-1}, t_i)$, such that $c((t' - \varepsilon, t' + \varepsilon)) \subseteq U$ and

$$\left\langle \xi^i \frac{\partial f}{\partial x^i}(c(t)), \frac{d^2(f \circ c)}{dt^2}(t) \right\rangle > 0$$

for all $t \in (t' - \varepsilon, t' + \varepsilon)$. Now let $\varphi : \mathbb{R} \longrightarrow \mathbb{R}$ be a bump function with

$$0 \leq \varphi \leq 1, \quad \operatorname{supp} \varphi \subseteq (t' - \varepsilon, t' + \varepsilon), \quad \text{and} \quad \varphi(t') = 1.$$

Set

$$h(s, t) = \begin{cases} x^{-1}\big(x(c(t)) + s\, \varphi(t)\, \xi\big) & \text{for } t \in (t' - \varepsilon, t' + \varepsilon), \\ c(t) & \text{otherwise.} \end{cases}$$

Then h is a proper variation of c. The variation field V of h satisfies

$$df(c(t))(V(t)) = \varphi(t)\, \xi^i \frac{\partial f}{\partial x^i}(c(t))$$

for all $t \in (t' - \varepsilon, t' + \varepsilon)$, and $V(t) = 0$ otherwise. However, the choice of v or ξ respectively implies $\delta L \neq 0$, which is a contradiction. Therefore, the tangential component of the second derivative of $f \circ c$ vanishes on the intervals $[t_{i-1}, t_i]$.

We next show that c is continuously differentiable, that is, that $\dot{c}(t_i-) = \dot{c}(t_i+)$ for all $0 < i < k$. To find a contradiction, we assume that $\Delta_i = \dot{c}(t_i-) - \dot{c}(t_i+) \neq 0$ for some i. We again choose a chart (U, x) on M, this time about $c(t_i)$, and a bump function φ as above, but now with support in $(t_i - \varepsilon, t_i + \varepsilon) \subseteq (t_{i-1}, t_{i+1})$ and with $\varphi(t_i) = 1$, such that $c((t_i - \varepsilon, t_i + \varepsilon)) \subseteq U$, and define a proper variation h of c as above, where ξ^i are the coefficients of Δ_i with respect to x. Since the tangential derivatives of $f \circ c$ are normal to f, it follows that $\delta L = \langle V(t_i), \Delta_i \rangle = \|\Delta_i\|^2 \neq 0$, a contradiction. So c is continuously differentiable.

It now only remains to show that c is smooth. For this, we refer to the discussion below: By the choice of t_i, c is already smooth on the intervals $[t_{i-1}, t_i]$. Now let $0 < i < k$ and (U, x) be a chart around $c(t_i)$. Let $\varepsilon > 0$ be chosen so that $c((t_i - \varepsilon, t_i + \varepsilon))$ is contained in U. Since the tangential component of the second derivative of $f \circ c$ vanishes on the intervals $(t_i - \varepsilon, t_i]$ and $[t_i, t_i + \varepsilon)$, the $c^j := x^j \circ c$ satisfy the second-order ordinary differential equations (4.40) on these intervals. Now, the values and the derivatives of the $c^j|_{(t_i - \varepsilon, t_i]}$ and $c^j|_{[t_i, t_i+\varepsilon)}$ agree at $t = t_i$, since c is continuously differentiable. Thus the c^j solve the equations in (4.40) on $(t_i - \varepsilon, t_i + \varepsilon)$. Since the coefficients Γ_{ij}^k of the equation are smooth, the c^j are smooth on these intervals. Therefore, c is smooth on $(t_i - \varepsilon, t_i + \varepsilon)$ and therefore on $[a, b]$.   □

Definition 4.2.14

We call a smooth curve $c: I \longrightarrow M$ a *geodesic*, if the tangential component of the second derivative of $f \circ c$ vanishes,

$$\left[\frac{d^2(f \circ c)}{dt^2}\right]^T = 0.$$

Example 4.2.15

Let $p, q \in M$. Suppose there is a curve $c_0: [a, b] \longrightarrow M$ such that $f \circ c_0$ is the constant-speed line segment in $\mathbb{R}^n$ from $f(p)$ to $f(q)$. Then c_0 is smooth and has constant speed $\|f(q) - f(p)\|/|b - a|$. The second derivative $f \circ c_0$ vanishes, so c_0 is a geodesic.

$f \circ c_0$ is the shortest among all the piecewise smooth curves in $\mathbb{R}^n$ from $f(p)$ to $f(q)$, and therefore, *a fortiori*, is the shortest among all curves of the form $f \circ c$. Thus, c_0 is a shortest curve connecting p to q in M. The first variation of arc length of every proper variation of c_0 must therefore vanish, so we again see that c_0 is a geodesic.

For the ruled surfaces of Example 4.2.3.4, the generators $t = \text{const}$ are lines in $\mathbb{R}^3$ and therefore are geodesics in the surface.   ∎

4.2.4 Covariant Derivative and Geodesics

The definitions of the length of piecewise smooth curves in M and of the geodesics as critical points of the length functional only involve the first fundamental form on M (resp. f). We therefore seek a formula for geodesics that only requires the first fundamental form. It is even advisable to aim higher. Let $X: I \longrightarrow TM$ be a smooth *vector field along c*, , i.e., $X(t) \in T_{c(t)}M$ for all $t \in I$. Then $Xf = df \circ X$ is a smooth tangent vector field along $f \circ c$. In general, $d(Xf)/dt$ is no longer tangent to f. The tangential part of $d(Xf)/dt$ is our focus. See also Lemma 4.2.4.

Definition 4.2.16

We call the unique vector field $X': I \longrightarrow TM$ along c with

$$df \circ X' = [d(Xf)/dt]^T$$

the *covariant derivative* of X, and we sometimes write $\nabla X/dt$ instead of X'. We call X *parallel* (along c) if $X' = 0$.

Geodesics are therefore, by Definition 4.2.14, characterized by the condition that the covariant derivative $\nabla \dot{c}/dt$ vanishes, or, in other words, that $\dot{c}$ is parallel along c.

Remark 4.2.17

1) To the best of my knowledge, we have Levi-Civita[9] to thank for the insight that the orthogonal projection onto the tangential component as in Definition 4.2.16 leads to a sensible kind of derivative: the covariant derivative.

2) For a vector field X along a curve c in M, X' is once again a vector field along c, while the usual derivative $\dot{X}$ of X takes values in TTM. The second derivative $\ddot{X}$ then has values in $TTTM$, and so forth. In contrast, higher covariant derivatives of X always remain vector fields along c.

3) We have already encountered the strategy of considering orthogonal projections of usual derivatives in our discussion of the normal fields of curves; see Definition 4.1.13 and the discussion which follows it.

We now discuss the computation of covariant derivatives in terms of local coordinates. Let (U, x) be a chart on M and $c: I \longrightarrow U$ a smooth curve. Let $X: I \longrightarrow M$ be a smooth vector field along c. Then

$$X = \xi^i \frac{\partial}{\partial x^i} \quad \text{and} \quad Xf = df \circ X = \xi^i \frac{\partial f}{\partial x^i} \quad \text{with} \quad \xi^j = \left\langle X, \frac{\partial}{\partial x^i} \right\rangle g^{ij},$$

where these equations are to be read as equations along c. With $c^i := x^i \circ c$ and since $f \circ c = (f \circ x^{-1}) \circ (x \circ c)$, we therefore obtain

$$\frac{d(Xf)}{dt} = \frac{d}{dt}\left(\xi^i \frac{\partial f}{\partial x^i}\right) = \dot{\xi}^i \frac{\partial f}{\partial x^i} + \dot{c}^j \xi^i \frac{\partial^2 f}{\partial x^j \partial x^i}. \tag{4.36}$$

Since the partial derivatives $\partial f/\partial x^i$ are tangent to f, the first term is already tangent to f. The second term is a linear combination of the second partial derivatives of f with respect to x, and is not, in general, tangent to f. The tangential components of the second partial derivatives are, however, pointwise linear combinations of the first partial derivatives, which are bases of their respective tangent spaces $T_p f$,

$$\left[\frac{\partial^2 f}{\partial x^i \partial x^j}\right]^T =: \Gamma_{ij}^k \frac{\partial f}{\partial x^k}. \tag{4.37}$$

The coefficients $\Gamma_{ij}^k : U \longrightarrow \mathbb{R}$ are called *Christoffel symbols*.[10] Using Definition 4.2.16 and (4.36), we can now compute the covariant derivative of X to be

$$\frac{\nabla X}{dt} = X' = \left(\dot{\xi}^k + \Gamma_{ij}^k \dot{c}^i \xi^j\right) \frac{\partial}{\partial x^k}. \tag{4.38}$$

---

[9] Tullio Levi-Civita (1873–1941).
[10] Elwin Bruno Christoffel (1829–1900).

where Γ_{ij}^k, strictly speaking, should be read as $\Gamma_{ij}^k \circ c$. By Lemma 4.2.4, or by Proposition 4.2.23 below, the Christoffel symbols are smooth. Therefore, the covariant derivative of X is also smooth. Since the second derivatives $\partial^2 f/\partial x^i \partial x^j$ are symmetric in i and j, the Christoffel symbols are symmetric in their two lower indices, $\Gamma_{ij}^k = \Gamma_{ji}^k$.

Proposition 4.2.18 *Let X and Y be smooth vector fields along c. Then the following hold:*
1. *(Linearity) For $\alpha, \beta \in \mathbb{R}$, $(\alpha X + \beta Y)' = \alpha X' + \beta Y'$.*
2. *(Product rule 1) For smooth $\varphi: I \longrightarrow \mathbb{R}$, $(\varphi X)' = \dot{\varphi} X + \varphi X'$.*
3. *(Product rule 2) The function $\langle X, Y \rangle$ has derivative $\langle X', Y \rangle + \langle X, Y' \rangle$.*

Proof
Claims 1 and 2 follow immediately from Definition 4.2.16 or from (4.38). By the definition of the first fundamental form $\langle X, Y \rangle = \langle Xf, Yf \rangle$, and therefore

$$d\langle X, Y \rangle/dt = d\langle Xf, Yf \rangle/dt = \langle d(Xf)/dt, Yf \rangle + \langle Xf, d(Yf)/dt \rangle$$
$$= \langle df \circ X', df \circ Y \rangle + \langle df \circ X, df \circ Y' \rangle = \langle X', Y \rangle + \langle X, Y' \rangle.$$

In passing from the first line to the second, we have used that, in the first term, the normal part of $d(Xf)/dt$ is by definition perpendicular to f and therefore is perpendicular to $Yf = df \circ Y$. We reason analogously for the second term. □

Corollary 4.2.19 *For parallel vector fields X and Y along c the following hold:*
1. *Linear combinations $\alpha X + \beta Y$ are themselves parallel along c.*
2. *The function $\langle X, Y \rangle$ is constant.* □

Let $W \subseteq \mathbb{R}^2 = \{(s, t) \mid s, t \in \mathbb{R}\}$ be an open subset, and let $\varphi : W \longrightarrow M$ be smooth. Then the partial derivatives $\partial\varphi/\partial s$ and $\partial\varphi/\partial t$ are vector fields along the s- and t-coordinate lines. We can therefore consider their covariant derivatives along these curves.

Proposition 4.2.20 *Let W be an open subset of the (s, t)-plane and let $\varphi : W \longrightarrow M$ be smooth. Then*

$$\frac{\nabla}{\partial s}\frac{\partial\varphi}{\partial t} = \frac{\nabla}{\partial t}\frac{\partial\varphi}{\partial s}.$$

Proof
By definition,

$$\frac{\nabla}{\partial s}\frac{\partial\varphi}{\partial t} = \left[\frac{\partial^2\varphi}{\partial s \partial t}\right]^T = \left[\frac{\partial^2\varphi}{\partial t \partial s}\right]^T = \frac{\nabla}{\partial t}\frac{\partial\varphi}{\partial s}.$$

□

With respect to local coordinates x and using (4.38), a vector field X along a smooth curve c is parallel, in the sense of Definition 4.2.16, if the coefficients ξ^i of X with respect to x solve the first-order ordinary differential equation

$$\dot{\xi}^k + \Gamma^k_{ij} \dot{c}^i \xi^j = 0. \tag{4.39}$$

The equation is linear in the ξ^i, so maximal solutions are defined on the entire domain of definition of the differential equation.

Corollary 4.2.21 *Let $c : I \longrightarrow M$ be a smooth curve and $t_0 \in I$. Then the following hold:*
1. *For $v \in T_{c(t_0)}M$ there is precisely one parallel vector field X along c with $X(t_0) = v$.*
2. *For a basis $(v_1, \ldots, v_m)$ of $T_{c(t_0)}M$ and parallel vector fields X_i along c with $X_i(t_0) = v_i$, $(X_1(t), \ldots, X_m(t))$ is a basis of $T_{c(t)}M$ for all $t \in I$. Furthermore, $\langle X_i(t), X_j(t) \rangle = \langle v_i, v_j \rangle$.* □

By definition, geodesics are smooth curves c with $\nabla \dot{c}/dt = 0$. Therefore, by (4.39), they are characterized as solutions to the second-order differential equation

$$\ddot{c}^k + \Gamma^k_{ij} \dot{c}^i \dot{c}^j = 0. \tag{4.40}$$

In contrast to (4.39), however, this equation is not linear in its solutions. Nonetheless, we can extract some important consequences from the fact that geodesics are solutions to an ordinary differential equation:

Corollary 4.2.22
1. *If $c_1 : I_1 \longrightarrow M$ and $c_2 : I_2 \longrightarrow M$ are geodesics with $I_1 \cap I_2 \neq \emptyset$ and $c_1|_{I_1 \cap I_2} = c_2|_{I_1 \cap I_2}$, then the concatenation $c : I_1 \cup I_2 \longrightarrow M$ of c_1 and c_2 is also a geodesic.*
2. *If c is a geodesic, so is $\tilde{c} = \tilde{c}(t) = c(at + b)$ for all $a, b \in \mathbb{R}$.*
3. *For $t \in \mathbb{R}$, $p \in M$, and $v \in T_p M$ there is precisely one maximal geodesic $c : I \longrightarrow M$ with $c(t) = p$ and $\dot{c}(t) = v$.* □

By *maximal*, we mean that the domain of every other geodesic with the given initial conditions is contained in I. In particular, I is open, since in local coordinates geodesics are solutions of the differential equation (4.40).

Proposition 4.2.23 *The Christoffel symbols associated to a chart x on M can be calculated from the coefficients of the first fundamental form in x. More explicitly,*

$$\Gamma^l_{ij} = \frac{1}{2} g^{kl} \left(\frac{\partial g_{jk}}{\partial x^i} + \frac{\partial g_{ik}}{\partial x^j} - \frac{\partial g_{ij}}{\partial x^k} \right).$$

Proof
We have

$$\frac{\partial g_{ij}}{\partial x^k} = \frac{\partial}{\partial x^k} \left\langle \frac{\partial f}{\partial x^i}, \frac{\partial f}{\partial x^j} \right\rangle = \left\langle \frac{\partial^2 f}{\partial x^k \partial x^i}, \frac{\partial f}{\partial x^j} \right\rangle + \left\langle \frac{\partial f}{\partial x^i}, \frac{\partial^2 f}{\partial x^k \partial x^j} \right\rangle.$$

Now $\partial f / \partial x^j$ and $\partial f / \partial x^i$ are tangent to f, so, on the right-hand side, only the tangential components of the second partial derivatives of f contribute. With (4.37), we then obtain

$$\frac{\partial g_{ij}}{\partial x^k} = \langle \Gamma^l_{ki} \frac{\partial f}{\partial x^l}, \frac{\partial f}{\partial x^j} \rangle + \langle \frac{\partial f}{\partial x^i}, \Gamma^l_{kj} \frac{\partial f}{\partial x^l} \rangle = \Gamma^l_{ki} g_{lj} + \Gamma^l_{kj} g_{il},$$

and similarly for $\partial g_{jk}/\partial x^i$ and $\partial g_{ik}/\partial x^j$. With the aid of the symmetry of the lower indices of the Christoffel symbols, we therefore obtain

$$\frac{\partial g_{jk}}{\partial x^i} + \frac{\partial g_{ik}}{\partial x^j} - \frac{\partial g_{ij}}{\partial x^k} = 2 \Gamma^l_{ij} g_{lk}.$$

Finally, therefore

$$2 \Gamma^l_{ij} = 2 \Gamma^\mu_{ij} \delta^l_\mu = 2 \Gamma^\mu_{ij} (g_{\mu k} g^{kl})$$

$$= 2 \left(\Gamma^\mu_{ij} g_{\mu k} \right) g^{kl} = g^{kl} \left(\frac{\partial g_{jk}}{\partial x^i} + \frac{\partial g_{ik}}{\partial x^j} - \frac{\partial g_{ij}}{\partial x^k} \right). \qquad \square$$

We see that the Christoffel symbols can be explicitly computed from the coefficients of the first fundamental form and their derivatives. Therefore, the Christoffel symbols are quantities belonging to interior geometry! With (4.39) and (4.40), geodesics and parallel vector fields are therefore also explicitly characterized as objects from interior geometry.

4.3 Exterior Geometry

The tangent space $T_p f$ approximates the image of f around p to first order, but does not describe, how that image bends in the ambient space. As in the case of curves, we therefore consider approximations of second order.

For $p \in M$, we consider the orthogonal decomposition $\mathbb{R}^n = T_p f \oplus N_p f$ with the corresponding projections π^T_p and π^N_p. Compare with (4.24) and (4.26). Then for the smooth map

$$x : M \longrightarrow T_p f, \quad x(q) = \pi^T_p (f(q) - f(p)), \tag{4.41}$$

we have $x(p) = 0$, and, since π^T_p is linear, $dx(p) = \pi^T_p \circ df(p) = df(p)$. Therefore $dx(p) : T_p M \longrightarrow T_p f$ is an isomorphism. The Inverse Function Theorem thus implies that there are open neighborhoods U of p in M and U' of 0 in $T_p f$, such that $x : U \longrightarrow U'$ is a diffeomorphism with $x(p) = 0$ and $dx(p) = df(p)$. Up to the choice of an isomorphism $T_p f \simeq \mathbb{R}^m$, (U, x) is thus the chart on M about p which best fits the region of M around p in the ambient space $\mathbb{R}^n$.

Proposition 4.3.1 (Local Normal Form) Let $h = \pi^N_p \circ (f - f(p)) \circ x^{-1} : U' \longrightarrow N_p f$ be the $N_p f$ component of $(f - f(p)) \circ x^{-1}$. Then $h(0) = 0$, $dh(0) = 0$, and

$$(f \circ x^{-1})(u) = f(p) + u + h(u) \quad \text{for all } u \in U'.$$

Proof

Since $x^{-1}(0) = p$, $h(0) = 0$. Furthermore,

$$dh(0) = \pi_p^N \circ df(p) \circ dx^{-1}(0).$$

Now $\operatorname{im} df(p) = T_p f$, so $\pi_p^N \circ df(p) = 0$. The rest is clear. $\qquad\square$

By Proposition 4.3.1, the Taylor expansion of h around 0 is given by

$$h(u) = \frac{1}{2} H_p(u, u) + \text{terms of third or higher order in } u, \tag{4.42}$$

where $H_p := D^2 h|_0$. Up to translation by $f(p)$ and terms of third or higher order in u,

$$Q_p = \{u + v \in \mathbb{R}^n \mid u \in T_p f, \ v = H_p(u, u)/2 \in N_p f\} \tag{4.43}$$

therefore describes the image of f around p. We call Q_p the *osculating paraboloid* to f at p. The osculating paraboloid to f at p is an m-dimensional submanifold of $\mathbb{R}^n$ and describes $f - f(p)$ about p to second order.

With the chart (U, x) on M around p as above, we now let $c \colon I \longrightarrow U$ be a smooth curve with $c(t_0) = p$ for some $t_0 \in I$. Then, by Proposition 4.3.1,

$$f \circ c = (f \circ x^{-1}) \circ (x \circ c) = f(p) + x \circ c + h \circ (x \circ c).$$

It therefore follows from (4.42) that

$$\frac{d^2(f \circ c)}{dt^2}(t_0) = \frac{d^2(x \circ c + h \circ (x \circ c))}{dt^2}(t_0)$$

$$= \frac{d^2(x \circ c)}{dt^2}(t_0) + H_p\big(dx(p)(\dot{c}(t_0)), dx(p)(\dot{c}(t_0))\big)$$

$$= \frac{d^2(x \circ c)}{dt^2}(t_0) + H_p\big(df(p)(\dot{c}(t_0)), df(p)(\dot{c}(t_0))\big).$$

The first term on the right is tangent to f at p, the second is normal. Therefore, the right-hand side is an orthogonal decomposition of the second derivative of $f \circ c$ at $t = t_0$. In this section, the second, normal, term is our focus.

Definition 4.3.2

We call the symmetric bilinear form

$$S_p : T_p M \times T_p M \longrightarrow N_p f, \quad S_p(v, w) := H_p(df(p)(v), df(p)(w)),$$

the *second fundamental form* of f at p.

We can summarize the computations above in the following equation:

$$\frac{d^2(f \circ c)}{dt^2} = df \circ \frac{\nabla \dot{c}}{dt} + S(\dot{c}, \dot{c}), \tag{4.44}$$

where $\nabla \dot{c}/dt$ denotes the covariant derivative of $\dot{c}$, see Definition 4.2.16. The reference points have been omitted in (4.44) for ease of reading. The first term is tangent to f, the second is normal. By (4.44), we can determine the second fundamental form S of f by polarization, without computing the special charts x or the local normal form of f as in Proposition 4.3.1.

Remark 4.3.3 In Definition 4.3.2 and (4.44) we again encounter the natural identification $df(p)\colon T_p M \longrightarrow T_p f$. In the case of a submanifold, this is the usual identification of $T_p M$ with a subspace of $\mathbb{R}^n$ and will therefore be omitted from our notation. For submanifolds, the above formulas simplify to

$$S_p = H_p \quad \text{and} \quad \ddot{c} = \frac{\nabla \dot{c}}{dt} + S(\dot{c}, \dot{c}), \tag{4.45}$$

where $\nabla \dot{c}/dt$ is the tangential part of $\ddot{c}$, and $S(\dot{c}, \dot{c})$ is the normal part.

Example 4.3.4

1) Let $M = I$ be an open interval and $c\colon I \longrightarrow \mathbb{R}^n$ be a regular curve, that is, an immersion. Then for $t \in I$, $T_t c = \mathbb{R} \cdot e(t)$ with $e(t) = \dot{c}(t)/\|\dot{c}(t)\|$ as in (4.1). The second fundamental form associates to $\partial/\partial t$ the normal component of $\ddot{c}$ (at the corresponding point), that is, $S(\partial/\partial t, \partial/\partial t) = \ddot{c} - \langle \ddot{c}, e \rangle e$. The length of these vectors is $\kappa \cdot \|\dot{c}\|^2$. Compare with (4.4).

2) Let $M = S_r^m \subseteq \mathbb{R}^{m+1}$ be the sphere with center 0 and radius r (and f be the inclusion). For $x \in S_r^m$, let $U = \{y \in S_r^m \mid \langle x, y \rangle > 0\}$. Then the orthogonal projection $\pi_x^T\colon U \longrightarrow T_x S_r^m = \{u \in \mathbb{R}^{m+1} \mid \langle x, u \rangle = 0\}$ is a diffeomorphism onto its image U' and is therefore a chart on S_r^m around x. Up to conflicts in the notation, it corresponds to the charts in (4.41) and Proposition 4.3.1, there denoted by x. As in Proposition 4.3.1 (for the case $f = $ inclusion), we can now write U as the graph of the map

$$h\colon U' \longrightarrow \mathbb{R} x = N_x S_r^m, \quad h(u) = \left(\sqrt{r^2 - \|u\|^2} - r\right)\frac{x}{r}.$$

Therefore, the second fundamental form of $M = S_r^m$ at x is given by

$$S_x(v, w) = -\frac{\langle v, w \rangle}{r^2} \cdot x.$$

With (4.44), we can determine $S_x(v, v)$ for a unit vector $v \in T_x S_r^n$ by determining the normal portion of the second derivative of a smooth curve c through x with derivative $\dot{c}(0) = v$. Here, the curve $c = c(t) = \cos(t)x + r\sin(t)v$ recommends itself, since $\ddot{c} = -c$ is normal to the sphere at all points, and therefore c is a geodesic. It thereby follows that $S_x(rv, rv) = -x$. Polarization then leads to the formula for S_x which we obtained in a different way above.

∎

Proposition 4.3.5 *For smooth vector fields* X, Y *on* M,

$$S(X, Y) = [XYf]^N.$$

Proof

Let $p \in M$ with the chart x around p as in (4.41) and the map h as in Proposition 4.3.1. Since $dh(0) = 0$, it then follows that

$$[XYf]^N (p) = [X(df \circ Y)]^N (p) = X(d(h \circ x) \circ Y)(p)$$
$$= D^2h(0)\big(df(p)(X(p)), df(p)(Y(p))\big) = S_p(X(p), Y(p)). \qquad \square$$

By (4.44), we already knew that we need not determine the special chart x as in (4.41) and h as in Proposition 4.3.1 to compute the second fundamental form. Proposition 4.3.5 provides another argument for this fact.

Now let $x : U \longrightarrow U'$ be an arbitrary chart on M. It then follows from Proposition 4.3.5 that the (vector-valued) entries in the fundamental matrix of the second fundamental form with respect to x are given by

$$h_{ij} := S\left(\frac{\partial}{\partial x^i}, \frac{\partial}{\partial x^j}\right) = \left[\frac{\partial^2 f}{\partial x^i \partial x^j}\right]^N. \qquad (4.46)$$

In conclusion, by (4.37) and (4.46), the tangential component of the second derivative of f determines the Christoffel symbols, and the normal component determines the coefficients of the second fundamental form.

4.3.1 Hypersurfaces

The classical case of submanifolds of Euclidean spaces are surfaces in $\mathbb{R}^3$. Since the discussion is essentially the same, we consider hypersurfaces $M \subseteq \mathbb{R}^{m+1}$ or immersions $f : M \longrightarrow \mathbb{R}^{m+1}$. Then, for all $p \in M$, $\dim N_p f = 1$. In particular, $N_p f$ contains precisely two unit vectors. For a given $p \in M$, let $n = n_p$ be one of these. Then we can write the second fundamental form S_p as

$$S_p(v, w) = S_p^n(v, w)\, n_p \quad \text{with} \quad S_p^n(v, w) := \langle S_p(v, w), n_p \rangle. \qquad (4.47)$$

We call S_p^n the *second fundamental form of* f *at* p *with respect to* n_p. If x is a chart on M about p, then we obtain, by (4.46),

$$h_{ij}^n(p) := S_p^n\left(\frac{\partial}{\partial x^i}(p), \frac{\partial}{\partial x^j}(p)\right) = \left\langle \frac{\partial^2 f}{\partial x^i \partial x^j}(p), n_p \right\rangle \qquad (4.48)$$

for the coefficients of the fundamental matrix of the second fundamental form. The sign of $h_{ij}^n(p)$ depends on the choice of the vector $n = n_p$.

The *Weingarten map*[11] is the self-adjoint endomorphism $L_p: T_pM \longrightarrow T_pM$ corresponding to the second fundamental form, which is characterized by

$$\langle L_p v, w \rangle = S_p^n(v, w) \quad \text{for all } v, w \in T_pM. \tag{4.49}$$

We know from linear algebra that the characteristic values and directions of a symmetric bilinear form on a Euclidean vector space correspond precisely to the eigenvalues and eigenvectors of the associated self-adjoint endomorphism of the vector space. Correspondingly, the characteristic values and directions of the second fundamental form are the eigenvalues and eigenvectors of the Weingarten map.

Definition 4.3.6

The characteristic values of the second fundamental form S_p^n are called the *principal curvatures* and the corresponding direction in T_pM *principal directions* of M or f at p. Regular curves $c: I \longrightarrow M$, such that $\dot{c}(t)$ is a principal direction for all $t \in I$ are called *lines of curvature*.

The principal directions do not depend on the choice of n_p, and only the signs of the principal curvatures do.

We now describe three classical results for surfaces in $\mathbb{R}^3$ connected with the correspondence between symmetric bilinear forms and self-adjoint endomorphisms:

1. The *Theorem of Rodrigues*[12] (1816) states that the minimum and the maximum of the function $S_p(v, v)$, where v runs through the unit vectors in T_pM, are attained at eigenvectors of L_p.

2. Let $E \subseteq \mathbb{R}^{m+1}$ be an affine plane through $f(p)$ tangent to n_p and a unit vector $u \in T_pf$. Furthermore, let $x : U \longrightarrow U'$ be the chart in (4.41). For a smooth curve $c: I \longrightarrow U$ through p, the image of $\sigma := f \circ c$ is contained in E if and only if $x \circ c$ lies in $\mathbb{R} \cdot u$. If this holds and $\dot{\sigma}(0)$ is a positive multiple of u, then we call σ a *normal section* of M (resp. f) through p in the direction u. See ◻ Fig. 4.5. With h as in Proposition 4.3.1, for example,

$$\sigma_u = \sigma_u(t) = (f \circ x^{-1})(tu) = f(p) + tu + h(tu) \tag{4.50}$$

is one such normal section. Since normal sections in the direction u only differ in their parameterizations locally around $t = 0$, their oriented curvatures (at least at $t = 0$), in the plane E oriented by (u, n_p), are independent of the choice of normal section and therefore define a function of the unit vector u, written as $\kappa_o(u)$. *Euler's Theorem* states that κ_o (as a function of u) is either constant or attains its maximum and minimum in directions perpendicular to one another. To see this, write $u =$

---

[11]Julius Weingarten (1836–1910).
[12]Benjamin Olinde Rodrigues (1794–1851).

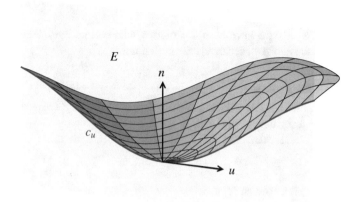

E

n

c_u

u

☐ **Fig. 4.5** A normal section

$df(p)(v)$ with $v \in T_p M$. With σ_u as above, we then have $\langle \dot{\sigma}_u(0), \ddot{\sigma}_u(0) \rangle = 0$ and therefore

$$\kappa_o(u) = \langle \ddot{\sigma}_u(0), n_p \rangle = S_p^n(v, v). \tag{4.51}$$

Therefore the maximum and minimum of κ_o correspond precisely to the extreme values of $S_p^n(v, v)$, where v runs through the unit vectors in $T_p M$. These will be attained in directions perpendicular to one another if the function $S_p^n = S_p^n(v, v)$ is not constant.

1) *Meusnier's Theorem*[13] (1776) complements Euler's Theorem: Let $v \in T_p M$ be a unit vector with $S_p^n(v, v) \neq 0$ and $c: I \longrightarrow M$ a smooth curve with $c(0) = p$ and $\dot{c}(0) = v$. Then for $\sigma := f \circ c, u := \dot{\sigma}(0) = df(p)(v)$ and $\ddot{\sigma}(0)$ are linearly independent. Let σ_u be a normal section of f through p in the direction u. Meusnier's Theorem now says that the orthogonal projection of the center of curvature M_u of σ_u at $t = 0$ onto the affine plane E through $f(p)$, spanned by $u = \dot{\sigma}(0)$ and $\ddot{\sigma}(0)$, is the center of curvature of σ at $t = 0$. Since $\dot{\sigma}(0)$ has norm 1, the curvature of σ at $t = 0$ is, by (4.4), given by $\|\ddot{\sigma}(0) - \langle \dot{\sigma}(0), \ddot{\sigma}(0) \rangle \dot{\sigma}(0)\|$. Therefore, Meusnier's Theorem follows from (4.44) and (4.51), since

$$\|\ddot{\sigma}(0) - \langle \dot{\sigma}(0), \ddot{\sigma}(0) \rangle \dot{\sigma}(0)\| \cos \theta = |\langle \ddot{\sigma}(0), n_p \rangle| = |S_p^n(v, v)|,$$

where $\theta \in [0, \pi/2]$ denotes the angle between E and $N_p f$. In particular, it follows from Meusnier's Theorem, that the osculating circles of all such $\sigma = f \circ c$ with

---

[13] Jean Baptiste Marie Charles Meusnier de la Place (1754–1793).

$\dot{c}(0) = v$ form a sphere with center M_u and radius $1/|S_p^n(v, v)|$, the so-called *Meusnier sphere*.

At every $p \in M$, we have a choice of two unit vectors $\pm n_p$. Naturally, it is disadvantageous to choose n_p separately for each point.

Definition 4.3.7

For an open subset $W \subseteq M$ a smooth map $n : W \longrightarrow S^m \subseteq \mathbb{R}^{m+1}$ is called a *Gauß map* on M (resp. f), if $n_p = n(p) \in N_p f$ for all $p \in W$.

Example 4.3.8

If $m = 2$ and (U, x) is a chart on M, then

$$\tilde{n} := \frac{\partial f}{\partial x^1} \times \frac{\partial f}{\partial x^2} \tag{4.52}$$

is perpendicular to f, and therefore $n := \tilde{n}/\|\tilde{n}\|$ is a Gauß map on U. For the case $m > 2$ there is an analogous formula. (Exercise: check the latter.) ∎

Remark 4.3.9 We call a connected hypersurface $M \subseteq \mathbb{R}^{m+1}$ *two-sided* if there is a global Gauß map $n : M \longrightarrow S^m$, otherwise we call M one-sided. Two-sidedness and one-sidedness are equivalent to the orientability and non-orientability of M respectively. Compare with ▶ Sect. 3.6. The Möbius strip is one-sided.

In the following, let $n \colon M \longrightarrow S^m$ be a Gauß map on f. For vector fields X and Y on M, we then have $\langle Yf, n \rangle = 0$, and therefore

$$0 = X\langle n, Yf \rangle = \langle n, XYf \rangle + \langle Xn, Yf \rangle.$$

By Proposition 4.3.5, $\langle XYf, n \rangle = S^n(X, Y)$. It therefore follows that

$$S^n(X, Y) = \langle n, XYf \rangle = -\langle Xn, Yf \rangle = -\langle df \circ Y, dn \circ X \rangle. \tag{4.53}$$

The second fundamental form thus agrees with the bilinear form $-\langle df, dn \rangle$, also written as $-df \cdot dn$. Now, for every $p \in M$, $T_p f$ is the orthogonal complement to $n(p) \in S^m$, so $T_p f = T_{n(p)} S^m$. Therefore the ranges of $df(p)$ and $dn(p)$ agree. Since $df(p) \colon T_p M \longrightarrow T_p f$ is by definition of the first fundamental form an orthogonal transformation, we can reformulate (4.53) as follows:

$$dn = -df \circ L, \tag{4.54}$$

where L denotes the Weingarten map. If, in particular, $M \subseteq \mathbb{R}^{m+1}$ and f is the inclusion, then $L = -dn$.

For a chart (U, x) on M, we write $L(\partial/\partial x^i) = a_i^l \, \partial/\partial x^l$ with as yet undetermined coefficients a_i^l. Since

$$h_{ij}^n = S^n\left(\frac{\partial}{\partial x^i}, \frac{\partial}{\partial x^j}\right) = \left\langle L\left(\frac{\partial}{\partial x^i}\right), \frac{\partial}{\partial x^j}\right\rangle = a_i^l g_{lj}$$

it follows that $a_i^k = a_i^l g_{lj} g^{jk} = h_{ij}^n g^{jk}$. Therefore, we finally obtain

$$L\left(\frac{\partial}{\partial x^i}\right) = h_{ij}^n g^{jk} \frac{\partial}{\partial x^k}. \qquad (4.55)$$

In other words, with respect to the basis $\partial/\partial x^i$, L is represented by the matrix with entries $h_{ij}^n g^{jk}$.

Up to sign, the elementary symmetric functions in the characteristic values of the second fundamental form correspond to the coefficients of the characteristic polynomial of the Weingarten map, and therefore, by (4.55), to the coefficients of the characteristic polynomial of the matrix $(h_{ij}^n g^{jk})$.

Definition 4.3.10

The determinant $K := \det L$ is called the *Gauß-Kronecker curvature*[14] of M (resp. f). The arithmetic mean $H := \operatorname{tr} L/m$ is called the *mean curvature* of M (resp. f).

If $\kappa_1, \ldots, \kappa_m$ denote the characteristic values of the second fundamental form S_p^n, that is, the eigenvalues of the Weingarten map L_p, then

$$K(p) = \kappa_1 \cdots \kappa_m \quad \text{and} \quad H(p) = (\kappa_1 + \cdots + \kappa_m)/m. \qquad (4.56)$$

The sign of H always depends on the choice of n_p, the sign of K does so only when m is odd.

We ought not neglect the important classical case:

Definition 4.3.11

If M is a surface, that is, $m = 2$, then

$$K := \det L = \det(h_{ij}^n)/\det(g_{ij})$$

is called the *Gaußian curvature* of M (resp. f). See ◻ Fig. 4.6.

---

[14]Leopold Kronecker (1823–1891).

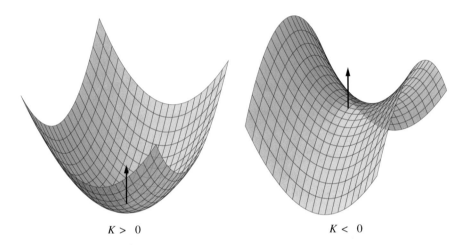

$K > 0$ $\qquad\qquad\qquad\qquad\qquad$ $K < 0$

■ Fig. 4.6 Positive and negative Gaußian curvature

Example 4.3.12

1) For the *surface of revolution* $f = f(t, \varphi)$ as in Example 4.2.3.2, we have

$$\tilde{n} = (-\dot{h} \cos(\varphi), -\dot{h} \sin(\varphi), \dot{r})$$

in the sense of Example 4.3.8, and $n = \tilde{n}/\|\tilde{n}\|$ is the associated Gauß map of f. The coefficients of the second fundamental form of f, as in (4.48), are

$$h^n_{tt} = \frac{\dot{r}\ddot{h} - \ddot{r}\dot{h}}{\sqrt{\dot{r}^2 + \dot{h}^2}}, \quad h^n_{t\varphi} = h^n_{\varphi t} = 0, \quad h^n_{\varphi\varphi} = \frac{r\dot{h}}{\sqrt{\dot{r}^2 + \dot{h}^2}}.$$

Since the first and second fundamental forms are both in diagonal form, meridians and lines of latitude are lines of curvature. The corresponding principal curvatures are

$$\kappa_t = \frac{\dot{r}\ddot{h} - \ddot{r}\dot{h}}{(\dot{r}^2 + \dot{h}^2)^{3/2}} \quad \text{and} \quad \kappa_\varphi = \frac{\dot{h}}{r\sqrt{\dot{r}^2 + \dot{h}^2}}.$$

If the profile curve is parameterized by arc length, then the Gaußian curvature is $K = -\ddot{r}/r$.

2) For the *generalized helicoid* $f = f(t, \varphi)$ as in Example 4.2.3.3,

$$\tilde{n} = (a\dot{x} \sin\varphi + a\dot{y} \cos\varphi, -a\dot{x} \cos\varphi + a\dot{y} \sin\varphi, x\dot{x} + y\dot{y})$$

in the sense of Example 4.3.8, and $n = \tilde{n}/\|\tilde{n}\|$ is the associated Gauß map of f. Therefore,

$$h^n_{tt} = a\frac{\dot{y}\ddot{x} - \dot{x}\ddot{y}}{\|\tilde{n}\|}, \quad h^n_{t\varphi} = h^n_{\varphi t} = -a\frac{\dot{x}^2 + \dot{y}^2}{\|\tilde{n}\|}, \quad h^n_{\varphi\varphi} = a\frac{\dot{x}y - x\dot{y}}{\|\tilde{n}\|}$$

with $\|\tilde{n}\|^2 = a^2(\dot{x}^2 + \dot{y}^2) + (x\dot{x} + y\dot{y})^2$. The formula for the Gaußian curvature is somewhat long, and so will not be written out here.

3) For the *ruled surface* $f = f(s, t)$ as in Example 4.2.3.4, $\tilde{n} = X \times (\dot{c} + s\dot{X})$ in the sense of Example 4.3.8, and $n = \tilde{n}/\|\tilde{n}\|$ is the associated Gauß map of f. Therefore,

$$h_{ss}^n = 0, \quad h_{st}^n = h_{ts}^n = \frac{\langle \dot{X}, X \times \dot{c} \rangle}{\|\tilde{n}\|}, \quad \text{and} \quad h_{tt}^n = \frac{\langle \ddot{c} + s\ddot{X}, X \times (\dot{c} + s\dot{X}) \rangle}{\|\tilde{n}\|}.$$

Since $h_{ss}^n = 0$, the Gaußian curvature K is non-positive.

∎

Definition 4.3.13

A point $p \in M$ is called an *umbilic point*, if the function $S_p^n(v, v)$ is constant, where v runs through the unit vectors in $T_p M$.

Clearly, $p \in M$ is an umbilic point if and only if the Weingarten map L_p is a multiple of the identity or, equivalently by (4.54), if $dn(p)$ is a multiple of $df(p)$.

Proposition 4.3.14 *If M is connected, $m \geq 2$, and all points in M are umbilic, then the image of f is contained in an affine hyperplane or a sphere.*

Proof

Since all points in M are umbilic, there is, by (4.54), a smooth function λ on M with $dn = \lambda\, df$. Now let (U, x) be a chart on M with U connected. Then $\partial n/\partial x^i = \lambda\, \partial f/\partial x^i$, and therefore

$$\frac{\partial \lambda}{\partial x^i} \frac{\partial f}{\partial x^j} + \lambda \frac{\partial^2 f}{\partial x^i \partial x^j} = \frac{\partial^2 n}{\partial x^i \partial x^j} = \frac{\partial^2 n}{\partial x^j \partial x^i} = \frac{\partial \lambda}{\partial x^j} \frac{\partial f}{\partial x^i} + \lambda \frac{\partial^2 f}{\partial x^j \partial x^i}.$$

Since the second terms on the left and right are equal, the first terms are as well. For $i \neq j$, however, $\partial f/\partial x^i$ and $\partial f/\partial x^j$ are linearly independent. Since $m \geq 2$, it therefore follows that the partial derivatives of λ vanish, so λ is constant on U.

For $\lambda = 0$ it follows that n is constant on U. Then the image of U under f is contained in an affine hyperplane of $\mathbb{R}^{m+1}$. For $\lambda \neq 0$ it follows that $f - n/\lambda = x_0 = $ constant on U. Since $\|n\| = 1$, we therefore obtain that $\|f - x_0\| = 1/|\lambda|$ on U. This shows the claim in the case $M = U$. But M is connected, so the claim follows. □

Affine planes in $\mathbb{R}^3$ and cylinders over regular plane curves are surfaces in space with vanishing Gaußian curvature. With an appropriate parameterization, one can (clearly) even ensure that their first fundamental forms have constant coefficients δ_{ij}. Consequently, the first fundamental form by no means determines the external appearance of a surface in space. Generalized helicoids provide an interesting example of this.

Example 4.3.15

Let $f = f(t, \varphi)$ be a generalized helicoid as in Example 4.2.3.3. We will show that—
up to a t-dependent shift in the φ-parameter (that is, a diffeomorphism of the parameter
domain)—there is a surface of revolution with the same first fundamental form. We first
consider surfaces of the form

$$\tilde{f}(t, \varphi) = f(t, \varphi + \alpha(t)).$$

The partial derivatives of $\tilde{f}$ are

$$\frac{\partial \tilde{f}}{\partial t}(t, \varphi) = \frac{\partial f}{\partial t}(t, \varphi + \alpha) + \frac{\partial f}{\partial \varphi}(t, \varphi + \alpha)\,\dot{\alpha},$$

$$\frac{\partial \tilde{f}}{\partial \varphi}(t, \varphi) = \frac{\partial f}{\partial \varphi}(t, \varphi + \alpha).$$

With x, y, and a as in Example 4.2.3.3, the coefficients of the first fundamental form of $\tilde{f}$
can be computed to be

$$\tilde{g}_{tt} = \dot{x}^2 + \dot{y}^2 + 2(x\dot{y} - \dot{x}y)\,\dot{\alpha} + (x^2 + y^2 + a^2)\,\dot{\alpha}^2,$$

$$\tilde{g}_{t\varphi} = \tilde{g}_{\varphi t} = x\dot{y} - \dot{x}y + (x^2 + y^2 + a^2)\,\dot{\alpha},$$

$$\tilde{g}_{\varphi\varphi} = x^2 + y^2 + a^2.$$

Now the first fundamental form of $\tilde{f}$ ought to also be that of a surface of revolution.
Therefore, we must have $\tilde{g}_{t\varphi} = \tilde{g}_{\varphi t} = 0$ and $\tilde{g}_{\varphi\varphi} = r^2$, where the profile curve of the surface
of revolution as in Example 4.2.3.2 is given by radius r and height h. So we must have

$$\dot{\alpha} = \frac{\dot{x}y - x\dot{y}}{r^2} \quad \text{with} \quad r = \sqrt{x^2 + y^2 + a^2}$$

and

$$\frac{(x\dot{x} + y\dot{y})^2}{r^2} + \dot{h}^2 = \dot{r}^2 + \dot{h}^2 = \tilde{g}_{tt} = \dot{x}^2 + \dot{y}^2 - \frac{(x\dot{y} - \dot{x}y)^2}{r^2}.$$

It therefore follows that

$$\dot{h} = \pm\frac{a}{r}\sqrt{\dot{x}^2 + \dot{y}^2}, \quad \text{so} \quad h = \pm\int \frac{a}{r}\sqrt{\dot{x}^2 + \dot{y}^2}.$$

We are free to choose the sign and constant of integration of h. Geometrically, these choices
correspond to a reflection or a translation in the z-direction respectively. The constant of
integration of α can also be chosen freely.

We can also assume that the profile curve of f is parameterized by arc length. Then

$$h = \pm\int \frac{a}{r}\,dt.$$

For the helicoid $f = (t \cos \varphi, t \sin \varphi, 1)$, we obtain (with appropriate choices)

$$r(t) = \sqrt{t^2 + 1} \quad \text{and} \quad h(t) = \operatorname{arcsinh} t.$$

So the corresponding surface of revolution, $t = \sinh h$ and $r = \cosh h$, is a catenoid. ∎

4.4 Gauß Equations and the Theorema Egregium

We now return to the covariant derivative.

Definition 4.4.1

For $Y \in V(M)$, the *covariant derivative* of Y in the direction $v \in T_p M$ is the unique tangent vector $\nabla_v Y \in T_p M$ with

$$df \circ \nabla_v Y = [v(Yf)]^T = [d(Yf)(v)]^T.$$

For $X \in V(M)$, the covariant derivative of Y in the direction of X is correspondingly given by $(\nabla_X Y)(p) := \nabla_{X(p)} Y = [(XYf)(p)]^T$.

Let $p \in M$ and (U, x) be a chart on M about p. Let $v = \xi^i \, \partial/\partial x^i(p)$ and $Y = \eta^i \, \partial/\partial x^i$ be a smooth vector field on U. Then $Yf = \eta^i \partial f/\partial x^i$ and so

$$v(Yf) = v(\eta^i) \frac{\partial f}{\partial x^i}(p) + \xi^i \eta^j (p) \frac{\partial^2 f}{\partial x^i \partial x^j}(p).$$

By the definition of the Christoffel symbols, therefore,

$$\nabla_v Y = \left(v(\eta^k) + \Gamma_{ij}^k(p) \xi^i \eta^j(p) \right) \frac{\partial}{\partial x^k}(p), \tag{4.57}$$

see (4.37). If X is a smooth vector field on U with $X = \xi^i \, \partial/\partial x^i$, then correspondingly

$$\nabla_X Y = \left(X\eta^k + \Gamma_{ij}^k \xi^i \eta^j \right) \frac{\partial}{\partial x^k}. \tag{4.58}$$

We see that $\nabla_X Y$ is again a smooth vector field.

Corollary 4.4.2
1. *If Y is a smooth vector field on M and $c: I \longrightarrow M$ is a smooth curve with $\dot{c}(t_0) =: v \in T_p M$, then $Y \circ c$ is a smooth vector field along c with $(Y \circ c)'(t_0) = \nabla_v Y$.*
2. *If X and Y are smooth vector fields on M, then*

$$XYf = [XYf]^T + [XYf]^N = df \circ \nabla_X Y + S(X, Y). \tag{4.59}$$

Proof
The first claim follows directly from (4.38) and (4.57), the second from Proposition 4.3.5 and Definition 4.4.1.   □

With (4.58), we see that the covariant derivative $\nabla_X Y$ with respect to a chart corresponds to the usual derivative of the coefficients of Y in the direction of X up to a correction term of order 0. Furthermore, Proposition 4.2.23 shows us that the covariant derivative belongs to the interior geometry of M.

Proposition 4.4.3 *For the map* $\nabla \colon \mathcal{V}(M) \times \mathcal{V}(M) \longrightarrow \mathcal{V}(M)$, $(X, Y) \mapsto \nabla_X Y$, *the following hold:*
1. ∇ *is bilinear, that is,* $\mathbb{R}$-*linear in both X and Y.*
2. *For all* $\varphi \in \mathcal{F}(M)$, $\nabla_{\varphi X} Y = \varphi \nabla_X Y$ *and* $\nabla_X(\varphi Y) = X(\varphi)Y + \varphi \nabla_X Y$.
3. ∇ *is symmetric:* $\nabla_X Y - \nabla_Y X = [X, Y]$.
4. ∇ *is metric (product rule):* $X \langle Y, Z \rangle = \langle \nabla_X Y, Z \rangle + \langle Y, \nabla_X Z \rangle$.

Proof
The first two assertions follow from (4.58). By the symmetry of the second fundamental form and (4.59),

$$df \circ [X, Y] = [X, Y]f = XYf - YXf$$
$$= df \circ \nabla_X Y - df \circ \nabla_Y X = df \circ (\nabla_X Y - \nabla_Y X).$$

Assertion 3 follows. The product rule follows from Proposition 4.2.18.3 together with Corollary 4.4.2.1.   □

A map $\nabla \colon \mathcal{V}(M) \times \mathcal{V}(M) \longrightarrow \mathcal{V}(M)$ that satisfies the first two conditions of Proposition 4.4.3 is called a *connection* or sometimes a *covariant derivative* on M. The special connection from Definition 4.4.1 is called the *Levi-Civita connection* of M (resp. f). It is distinguished among all connections on M by the last two conditions in Proposition 4.4.3.

We next discuss the *Riemann curvature tensor*,[15]

$$R(X, Y)Z := \nabla_X \nabla_Y Z - \nabla_Y \nabla_X Z - \nabla_{[X,Y]} Z, \tag{4.60}$$

a further object of interior geometry. Patient computation yields the following formula for vector fields X, Y, Z on the domain U of a chart x on M:

$$R(X, Y)Z = \left\{ \partial \Gamma_{jk}^l / \partial x^i - \partial \Gamma_{ik}^l / \partial x^j + \Gamma_{iv}^l \Gamma_{jk}^v - \Gamma_{jv}^l \Gamma_{ik}^v \right\} \xi^i \eta^j \zeta^k \frac{\partial}{\partial x^l}. \tag{4.61}$$

We see that R is pointwise linear in each of the coefficients ξ^i, η^j, and ζ^k of X, Y, and Z, and no derivatives of these coefficients appear. This justifies the designation of R as a tensor.

---

[15]Georg Friedrich Bernhard Riemann (1826–1866).

Proposition 4.4.4 (Gauß Equations) *For vector fields* X, Y, V, W *on* M,

$$\langle R(X, Y)V, W \rangle = \langle S(X, W), S(Y, V) \rangle - \langle S(X, V), S(Y, W) \rangle.$$

Proof

Since Vf is a map to the vector space $\mathbb{R}^n$,

$$XYVf - YXVf = [X, Y]Vf.$$

With (4.59) and Proposition 4.4.3.4, we obtain

$$
\begin{aligned}
\langle XYVf, Wf \rangle &= X\langle YVf, Wf \rangle - \langle YVf, XWf \rangle \\
&= X\langle [YVf]^T, Wf \rangle - \langle [YVf]^T, [XWf]^T \rangle - \langle [YVf]^N, [XWf]^N \rangle \\
&= X\langle \nabla_X Y, W \rangle - \langle \nabla_Y V, \nabla_X W \rangle - \langle S(Y, V), S(X, W) \rangle \\
&= \langle \nabla_X \nabla_Y V, W \rangle - \langle S(Y, V), S(X, W) \rangle,
\end{aligned}
$$

and analogously, $\langle YXVf, Wf \rangle = \langle \nabla_Y \nabla_X V, W \rangle - \langle S(X, V), S(Y, W) \rangle$. Finally, by (4.59), it further holds that

$$\langle [X, Y]Vf, Wf \rangle = \langle df \circ \nabla_{[X,Y]}V + S([X, Y], V), Wf \rangle = \langle \nabla_{[X,Y]}V, W \rangle.$$

Altogether, we obtain

$$
\begin{aligned}
0 &= \langle XYVf - YXVf - [X, Y]Vf, Wf \rangle \\
&= \langle \nabla_X \nabla_Y V - \nabla_Y \nabla_X V - \nabla_{[X,Y]}V, W \rangle \\
&\quad - \langle S(Y, V), S(X, W) \rangle + \langle S(X, V), S(Y, W) \rangle \\
&= \langle R(X, Y)V, W \rangle - \langle S(Y, V), S(X, W) \rangle + \langle S(X, V), S(Y, W) \rangle. \qquad \square
\end{aligned}
$$

Proposition 4.4.4 contains, in a more general form, one of the fundamental geometric insights of Gauß, namely, that the Gaußian curvature, defined with the aid of the second fundamental form, is a quantity belonging to interior geometry. In our formulation, Gauß's insight[16] reads as follows:

Theorema Egregium 4.4.5 (Gauß (1827)) *Let* M *be a surface and* $f: M \longrightarrow \mathbb{R}^3$ *be an immersion. Let* $p \in M$, *and let* $v, w \in T_p M$ *be linearly independent. Then*

$$K(p) = \frac{\langle R(v, w)w, v \rangle}{\|v^2\|\|w\|^2 - \langle v, w \rangle^2}.$$

---

[16]Gauß called this insight the *Theorema Egregium*, meaning "remarkable theorem". The word "egregium" is related etymologically to the word "egregious".

Proof

By Proposition 4.4.4, it follows that $\langle R(v, w)w, v \rangle = S_p^n(v, v)S_p^n(w, w) - S_p^n(v, w)^2$, independently of the choice of a normal vector n_p at p. Therefore, the right-hand side of the asserted equality becomes

$$\frac{S_p^n(v, v)S_p^n(w, w) - S_p^n(v, w)^2}{\|v\|^2\|w\|^2 - \langle v, w \rangle^2}.$$

The number does not depend on the choice of v and w, as long as v and w are linearly independent. We can, for example, choose a chart x around p, and set $v = (\partial/\partial x^1)(p)$ and $w = (\partial/\partial x^2)(p)$. Then

$$\frac{S_p^n(v, v)S_p^n(w, w) - S_p^n(v, w)^2}{\|v^2\|\|w\|^2 - \langle v, w \rangle^2} = \frac{h_{11}^n h_{22}^n - (h_{12}^n)^2}{g_{11}g_{22} - g_{12}^2}(p)$$

$$= \det(h_{ik}^n g^{kj})(p) = K(p). \qquad \square$$

More generally, the terms on the right-hand side of the equation in Proposition 4.4.4 coming from the second fundamental form are quantities belonging to interior geometry, since the Riemann curvature tensor is. Among these quantities is the *sectional curvature*, introduced by Riemann, that associates a curvature to tangent planes to M: if $p \in M$, P is a linear 2-dimensional subspace of T_pM, and (v, w) is a basis of P, then

$$K(P) := \frac{\langle R(v, w)w, v \rangle}{\|v^2\|\|w\|^2 - \langle v, w \rangle^2} \qquad (4.62)$$

is the sectional curvature of P. This generalizes the Gaußian curvature and was introduced by Riemann in his famous Habilitation[17] lecture (1854) before he had defined his curvature tensor.

In the following examples we replace $\mathbb{R}^n$ by a general finite-dimensional Euclidean vector space W. By means of an orthonormal transformation, we can then identify W with $\mathbb{R}^{\dim W}$ and, accordingly, use the previous concepts and results with W as our ambient space.

Example 4.4.6

Let $\mathbb{K} \in \{\mathbb{R}, \mathbb{C}, \mathbb{H}\}$ and V be an n-dimensional vector space over $\mathbb{K}$ together with a positive definite sesquilinear form, denoted by (v, w). After a choice of an orthonormal basis, $V \cong \mathbb{K}^n$ with sesquilinear form $(x, y) = \sum \bar{x}_i y_i$.

1) Let $G = \{A \in \mathrm{End}(V) \mid A^*A = \mathrm{id}\}$ be the Lie group of all endomorphisms which preserve the sesquilinear form on V. In Example 2.3.8.3 we saw, modulo the

---

[17]In the German-speaking world, a Habilitation is a post-doctoral qualification necessary to teach at the university level.

identification of V with $\mathbb{K}^n$ seen as a Euclidean space, that G is a submanifold of $W = \mathrm{End}(V)$ with

$$T_E G = \{C \in \mathrm{End}(V) \mid C^* = -C\}. \tag{4.63}$$

The real part of the sesquilinear form $(A, B) := \mathrm{tr}(A^* B)$ is a scalar product on $\mathrm{End}(V) \cong \mathbb{R}^{dn \times n}$, $d = \dim_\mathbb{R} \mathbb{K}$. The corresponding orthogonal complement of $T_E G$ in $\mathrm{End}(V)$ is

$$N_E G = H(V) = \{C \in \mathrm{End}(V) \mid C^* = C\}. \tag{4.64}$$

For all $A \in G$, $A T_E G A^{-1} = T_E G$ and $A H(V) A^{-1} = H(V)$, and left and right translation by A preserve the sesquilinear form on $\mathrm{End}(V)$:

$$(AC, AD) = (CA, DA) = (C, D) = \mathrm{tr}(C^* D)$$

for all $C, D \in \mathrm{End}(V)$. In particular,

$$T_A G = A T_E G = T_E G A \quad \text{and} \quad N_A G = A H(V) = H(V) A. \tag{4.65}$$

For all $A \in G$ and $C \in \mathrm{End}(V)$, it therefore follows that

$$\pi_A^T(AC) = \frac{1}{2} A(C - C^*) \quad \text{and} \quad \pi_A^N(AC) = \frac{1}{2} A(C + C^*). \tag{4.66}$$

The (tangential) left-invariant vector fields on G are of the form $X_C(A) = AC$ with $C \in T_E G$, compare with Exercise 2.7.14.1. Since X_C is the restriction of right multiplication by C to G, and the latter is linear on $\mathrm{End}(V)$,

$$dX_C(A)(AD) = ADC = X_{DC}(A)$$

for all $A \in G$ and $D \in T_E G$. For left invariant vector fields X_C and X_D (with $C, D \in T_E G$), we can use (4.59) to show that

$$dX_D(X_C) = \nabla_{X_C} X_D + S(X_C, X_D) = X_{(CD-DC)/2} + X_{(CD+DC)/2} \tag{4.67}$$

for the covariant derivative and second fundamental form of G. For $B, C, D \in T_E G$, the curvature tensor R of G then becomes

$$R(X_B, X_C)X_D = X_F \quad \text{with} \quad F = -\frac{1}{4}[[B, C], D] \in T_E G. \tag{4.68}$$

Since $B^* = -B$ and $[B, C]^* = -[B, C]$, we therefore obtain

$$\langle R(X_B, X_C)X_C, X_B \rangle = \frac{1}{4} \mathrm{Re} \, \mathrm{tr}([[B, C], C]B) = \frac{1}{4} \|[B, C]\|^2 \tag{4.69}$$

for the sectional curvature of G. In particular, the sectional curvature is non-negative.

Now let $A = A(t) = Be^{tC}$, with $B \in G$, $C \in T_E G$, and let the exponential series be as in Exercise 2.7.6. Then A is a smooth curve in G with

$$\dot{A} = AC \quad \text{and} \quad \ddot{A} = AC^2. \tag{4.70}$$

Now $C^2 \in N_E G$, so $\ddot{A} \in N_A G$. Therefore, A is a geodesic, and so we have characterized all the geodesics of G.

2) Let $G_k(V)$ be the Grassmannian manifold of k-dimensional $\mathbb{K}$-linear subspaces in V as in Example 2.1.15.5. For $P \in G_k(V)$, let $F(P): V \longrightarrow V$ be the orthogonal projection of V onto P with image P and kernel $P^\perp$. If $(e_1, \ldots, e_k)$ is an orthonormal basis of P, then

$$F(P)(v) = \sum_{1 \leq i \leq k} e_i \, (e_i, v). \tag{4.71}$$

We now extend $(e_1, \ldots, e_k)$ to an orthonormal basis $E = (e_1, \ldots, e_n)$ of V and consider the chart κ_E on $G_k(V)$ around P associated to E as in Example 2.1.15.5. Then for $A = (a_i^\mu) \in \mathbb{K}^{n-k,k}$, $\kappa_E^{-1}(A)$ is the graph of the linear map $P \longrightarrow P^\perp$, $e_i u^i \mapsto e_{k+j} a_i^j u^i$. Therefore,

$$F(\kappa_E^{-1}(A)) = \begin{pmatrix} 1 & 0 \\ A & 0 \end{pmatrix} \begin{pmatrix} 1 & -A^* \\ A & 1 \end{pmatrix}^{-1} \tag{4.72}$$

is the matrix of $F(\kappa_E^{-1}(A))$ with respect to E. In particular, it therefore follows that F is smooth with differential

$$dF(P)(A) = \begin{pmatrix} 0 & A^* \\ A & 0 \end{pmatrix},$$

where $T_P G_k(V)$ is identified with $\mathrm{Hom}(V, V^\perp) \cong \mathbb{K}^{n-k,k}$ as in Example 2.2.3.3. In particular, $F: G_k(V) \longrightarrow \mathrm{End}(V)$ is an embedding.

We equip $\mathrm{End}\, V$ with the real part of the sesquilinear form $\mathrm{tr}(X^*Y)$ as a scalar product. Then the first fundamental form of F in P becomes

$$\langle A, B \rangle_P = \langle dF(P)(A), dF(P)(B) \rangle$$

$$= \mathrm{Re}(\mathrm{tr}(A^*B) + \mathrm{tr}(AB^*)) = 2\,\mathrm{Re}\,\mathrm{tr}(A^*B).$$

To determine the geodesics and the second fundamental form, we now write P_0 in place of P. Again let $A \in \mathbb{K}^{n-k,k}$ and

$$P = P(t) = e^{tB} P_0 \quad \text{with} \quad B = \begin{pmatrix} 0 & -A^* \\ A & 0 \end{pmatrix}. \tag{4.73}$$

Then P is a smooth curve in $G_k(V)$ with $P(0) = P_0$ and $\dot{P}(0) = A$. Now $(e^{tB})^* = e^{-tB}$, so the e^{tB} preserve the sesquilinear form $(.,.)$ on V. It therefore follows that

$$F(P(t)) = e^{tB} F(P_0) e^{-tB} = e^{tB} \begin{pmatrix} 1 & 0 \\ 0 & 0 \end{pmatrix} e^{-tB} \qquad (4.74)$$

and $T_P F = e^{tB} T_{P_0} F e^{-tB}$. The second derivative of $F \circ P$ is computed to be

$$\frac{d^2(F \circ P)}{dt^2} = e^{tB} \begin{pmatrix} -2A^*A & 0 \\ 0 & 2AA^* \end{pmatrix} e^{-tB}. \qquad (4.75)$$

Therefore, the second derivative of $F \circ P$ is normal to F, so $P = P(t)$ is a geodesic, and so we have characterized all the geodesics of $G_k(V)$. By polarization, we furthermore obtain the second fundamental form F at P_0,

$$S_{P_0}(A, B) = \begin{pmatrix} -A^*B - B^*A & 0 \\ 0 & AB^* + BA^* \end{pmatrix}.$$

Since the $F(P)$ are orthogonal projections onto k-dimensional subspaces of V, $F(P)^* = F(P)$ and $\mathrm{tr}(F(P)) = k$. The image of F therefore lies in a sphere of radius $\sqrt{k}$ in the space of self-adjoint endomorphisms $W = H(V) \subseteq \mathrm{End}(V)$. Thus, the $T_P F$ and the images of the second fundamental forms S_P of F are contained in $H(V)$, and our formulas above show this explicitly.

∎

The curvature tensor and sectional curvature are quantities belonging to interior geometry. This is the beginning of Riemannian geometry based on Riemannian metrics, that is, smooth families g_p of scalar products on the $T_p M$, which are no longer required to be the first fundamental form of a given immersion $M \longrightarrow \mathbb{R}^n$. As in the case of the first fundamental form, one obtains a Levi-Civita connection, a curvature tensor, and the associated sectional curvature, and, armed with these, enters the realm of Riemannian geometry.

4.5 Supplementary Literature

Good sources to round off the discussion in this chapter are [Kl, dC], [Sp2, chapters 1&2], [Sp3, Ho] and [ST]. The more classical [MP] and [St1, St2, St3] are also interesting. In [Ch], introductory articles on many aspects of global differential geometry can be found. Beside these, there are also a number of newer introductions to differential geometry, for example [Bä, EJ, Le], and [Kü].

4.6   Exercises

Exercise 4.6.1
1. The length of a smooth curve $c: [a, b] \longrightarrow \mathbb{R}^n$ is invariant under monotone reparameterization: If $\varphi: [\alpha, \beta] \longrightarrow [a, b]$ is monotone, surjective, and smooth, then $L(c \circ \varphi) = L(c)$.
2. Show that analogues of the statements about length and energy in the corresponding part of ▶ Sect. 4.1 hold for piecewise smooth curves (as in (3.4)).
3. Let $c: [a, b] \longrightarrow \mathbb{R}^n$ be a piecewise smooth curve. For subdivisions

$$U: a = t_0 < \cdots < t_k = b$$

of $[a, b]$, let $\delta(U) = \max(t_i - t_{i-1})$ and $L(U)$ be the length of the piecewise linear curve with corners at $c(t_0), c(t_1), \ldots, c(t_k)$. Show: if (U_n) is a sequence of subdivisions of $[a, b]$ with $\delta(U_n) \longrightarrow 0$, then $L(U_n) \longrightarrow L(c)$.
4. Let $c: [0, \infty) \longrightarrow \mathbb{R}^2$ be a smooth curve of the form $c(t) = (t, y(t))$. If $\dot{y}(t)$ converges for $t \to \infty$, then $\lim_{t \to \infty} L(c|_{[0,t]}) / \|(t, y(t)) - (0, y(0))\| = 1$.   ∎

Exercise 4.6.2
Compute the speed and curvature of the curves $c: \mathbb{R} \longrightarrow \mathbb{R}^2$ given by
1. $c(t) = (t, t^k)$ with $k \geq 0$ and, more generally, $c(t) = (t, f(t))$;
2. $c(t) = \exp(t)(\cos t, \sin t)$ and, more generally, $c(t) = f(t)(\cos t, \sin t)$ with $f > 0$;
3. $c(t) = (a \cos t, b \sin t)$.
Sketch each curve or curve type.   ∎

Exercise 4.6.3
Let $c: I \longrightarrow \mathbb{R}^n$ be a regular curve with curvature κ.
1. If the image of c lies in a circle K of radius R, then, for all $t \in I$, K is the osculating circle of c and $\kappa \equiv 1/R$.
2. The image of c lies in a line if and only if κ vanishes.
3. For $x \in \mathbb{R}^n$, suppose the function $r = r(t) = \|c(t) - x\|$ has a relative maximum at $t_0 \in I$. Then $\kappa(t_0) = 1/R(t_0) \geq 1/r(t_0)$.
4. If $\varphi: J \longrightarrow I$ is a change of parameters, then $\tilde{\kappa} = \kappa \circ \varphi$, where $\tilde{\kappa}$ denotes the curvature of the curve $\tilde{c} := c \circ \varphi$.
5. If $B : \mathbb{R}^n \longrightarrow \mathbb{R}^n$ is a motion, then the curve $B \circ c$ has the same speed and curvature as c.   ∎

Exercise 4.6.4
Let $c: I \longrightarrow \mathbb{R}^2$ be a regular plane curve.
1. If $\varphi: J \longrightarrow I$ is a change of parameters with $\dot{\varphi} > 0$, then $\tilde{\kappa}_o = \kappa_o \circ \varphi$, where $\tilde{\kappa}_o$ denotes the oriented curvature of $\tilde{c} := c \circ \varphi$.
2. If $B : \mathbb{R}^2 \longrightarrow \mathbb{R}^2$ is an orientation-preserving motion, then the regular plane curve $B \circ c$ has the same oriented curvature as c.   ∎

Exercise 4.6.5

In the following, let $c\colon I \longrightarrow \mathbb{R}^2$ be a regular plane curve with field of directions e, principal normal field n and oriented curvature κ_o.

1. If $c + n$ is constant, then c moves along a circle of radius 1.
2. If $\kappa_o(t) \neq 0$ for all $t \in I$, then the curve $a = a(t) = c(t) + n(t)/\kappa_o(t)$ sending t to the center of curvature of c at t is called the *evolute* of c. Compute e, n, κ_o, and a for the parabola (t, t^2) and the catenary $(t, \cosh t)$.
3. Let $\kappa_o(t) \neq 0$ for all $t \in I$ and a be the evolute of c. Show that $\dot{a}(t)$ is a multiple of $n(t)$ for all $t \in I$ and conclude that, for all points where it is defined, the tangent to the evolute intersects the curve c perpendicularly at $c(t)$. For $s < t$ in I, compute the length of the arc $a|_{[s,t]}$ of the evolute, and compare it with the radius of curvature of c at s and t.
4. Again let $\kappa_o(t) \neq 0$ for all $t \in I$ and c be parameterized by arc length for simplicity. For $\beta \in \mathbb{R} \setminus I$, we then call $b = b(t) = c(t) + (\beta - t)\dot{c}(t)$, $t \in I$, the *evolvent* of c. One can image b as the endpoint of a thread wound tautly around c. Compute the field of directions, the principal normal field, and the curvature of b, and conclude that the curve c is the evolute of its evolvent.

■

Exercise 4.6.6

Compute the field of directions, principal normal field, and binormal field of the *helix* $c\colon \mathbb{R} \longrightarrow \mathbb{R}^3$, $c(t) = (r \cos t, r \sin t, ht)$ with $r, h \in \mathbb{R}, r > 0$. Show that the speed, curvature, and torsion of the helix are given by

$$\|\dot{c}(t)\| = \sqrt{r^2 + h^2}, \quad \kappa(t) = \frac{r}{r^2 + h^2}, \quad \text{and} \quad \tau(t) = \frac{h}{r^2 + h^2},$$

and are therefore, in particular, constant. Up to reparameterization and motions we therefore obtain all space curves with constant curvature $\kappa > 0$ and torsion τ. Also verify that the helix is a *generalized helix*, that is, that $\dot{c}$ and the z-axis enclose a constant angle.

■

Exercise 4.6.7

Let $c\colon I \longrightarrow \mathbb{R}^3$ be a smooth space curve such that $\dot{c}(t)$ and $\ddot{c}(t)$ are linearly independent for all $t \in I$.

1. The curvature and torsion of c are given by

$$\kappa(t) = \frac{\|\dot{c}(t) \times \ddot{c}(t)\|}{\|\dot{c}(t)\|^3} \quad \text{and} \quad \tau(t) = \frac{\det(\dot{c}(t), \ddot{c}(t), \dddot{c}(t))}{\|\dot{c}(t) \times \ddot{c}(t)\|^2}.$$

2. The image of c lies in an affine plane in $\mathbb{R}^3$ if and only if the torsion of c vanishes.
3. If c is parameterized by arc length and $t_0 = 0 \in I$, then

$$c(t) = c(0) + \left(t - \frac{t^3}{6}\kappa_0^2\right)e_0 + \left(\frac{t^2}{2}\kappa_0 + \frac{t^3}{6}\dot{\kappa}_0\right)n_0 + \frac{t^3}{6}\kappa_0\tau_0 b_0 + o(t^3),$$

where in each case the index 0 denotes the value at 0. Sketch the projections of c onto the *osculating*, *normal*, and *rectifying planes* of c at $t = 0$, that is, the planes spanned by the pairs of vectors (e_0, n_0), (n_0, b_0), and (e_0, b_0) respectively.

■

Exercise 4.6.8 (Concerning Definition 4.1.13)
1. Determine the parallel normal fields along a regular space curve with vanishing torsion.
2. Determine the parallel normal fields along the helix; see Example 4.1.18 and Exercise 4.6.6.   ∎

Exercise 4.6.9
1. Write $S^2 \setminus \{(0, 0, \pm 1)\}$ as a surface of revolution.
2. Write the one-sheeted *hyperboloid* $x^2 + y^2 - z^2 = 1$ as a surface of revolution and—in two different ways—as a ruled surface.
3. Represent the Möbius strip as a ruled surface.   ∎

Exercise 4.6.10
Let $c = (r, h) : I \longrightarrow \mathbb{R}^2$ be the profile curve of a surface of revolution $f = f(t, \varphi)$ as in Example 4.2.3.2. Show that one can reparameterize c so that
(a) $g_{tt} \equiv 1$, or
(b) $g_{tt} = g_{\varphi\varphi}$, or
(c) $g_{tt} g_{\varphi\varphi} = 1$.
In each case, we have $g_{t\varphi} = g_{\varphi t} = 0$. Therefore f (or more precisely $df(t, \varphi)$ for all $(t, \varphi) \in I \times \mathbb{R}$) is *angle-preserving* or *conformal* in case b), and is *area-preserving* in case c).   ∎

Exercise 4.6.11
The *gradient* of a smooth function φ on M is the vector field $\operatorname{grad} \varphi$, such that $\langle \operatorname{grad} \varphi(p), v \rangle = d\varphi(p)(v)$ for all p in M and $v \in T_p M$. Show: With respect to a chart (U, x) on M,

$$\operatorname{grad} \varphi = g^{ij} \frac{\partial \varphi}{\partial x^i} \frac{\partial}{\partial x^j} \quad \text{on } U. \tag{4.76}$$

∎

Exercise 4.6.12
1. For a geodesic c, $\|\dot{c}\|$ is constant.
2. Up to reparameterization, the meridians of surfaces of revolution are geodesics. Under what conditions on the profile curve are they geodesics? Which latitudes are geodesics?
3. For $x, y \in \mathbb{R}^{m+1}$ such that $\|x\| = \|y\| = 1$ and $\langle x, y \rangle = 0$, the curve $c(t) = r\cos(t)x + r\sin(t)y$, $t \in \mathbb{R}$, is a geodesic on the sphere S_r^m as in Example 4.2.1.2. Up to reparameterization, every geodesic on S_r^m is of this form.
4. Let $c : I \longrightarrow \mathbb{R}^3$ be a space curve such that $\dot{c}$ and $\ddot{c}$ are pointwise linearly independent, and let $f = f(s, t) = c(t) + s b(t)$ be the ruled surface spanned by c and the binormal field b of c. Show: f is an immersion of $M = \mathbb{R} \times I$, and the curve $(t, 0)$, $t \in I$, is a geodesic in M.   ∎

Exercise 4.6.13

1. For a regular curve $c: I \longrightarrow M$ with field of directions $e := \dot{c}/\|\dot{c}\|$ in M, we call $\|\nabla \dot{c}/dt - \langle e, \nabla \dot{c}/dt\rangle e\|/\|\dot{c}\|^2$ the *geodesic curvature* of c; compare with (4.4). Show: Up to reparameterization c is a geodesic if and only if the geodesic curvature of c vanishes.

2. Let $f: M \longrightarrow \mathbb{R}^n$ be an immersion of the surface M and $c: I \longrightarrow M$ be a regular curve. Let n be one of the two vector fields along c with $\langle \dot{c}, n\rangle = 0$ and constant norm 1. The corresponding oriented geodesic curvature of c is then $\kappa_o := \langle \nabla \dot{c}/dt, n\rangle/\|\dot{c}\|^2$ with $e = \dot{c}/\|\dot{c}\|$. Prove that

$$\nabla e/dt = \|\dot{c}\|\kappa_o n \quad \text{and} \quad \nabla n/dt = -\|\dot{c}\|\kappa_o e.$$

3. Determine the oriented geodesic curvature of the latitudes of a surface of revolution as in Example 4.2.3.2. Discuss the special case of spheres. ∎

Exercise 4.6.14

1. Let $M \subseteq \mathbb{R}^n$ be a linear subspace and let $c: I \longrightarrow M$ be a smooth curve. Show: A vector field $X: I \longrightarrow M$ along c is parallel in the sense of Definition 4.2.16 if and only if it is parallel in the usual sense.

2. Verify with the use of (4.37) that, with respect to the coordinates (t, φ), the Christoffel symbols of the surface of revolution with profile curve $c = (r, h)$ are given by

$$\Gamma_{tt}^t = \frac{\dot{r}\ddot{r} + \dot{h}\ddot{h}}{\|\dot{c}\|^2}, \quad \Gamma_{\varphi\varphi}^t = -\frac{r\dot{r}}{\|\dot{c}\|^2}, \quad \Gamma_{\varphi t}^\varphi = \Gamma_{t\varphi}^\varphi = \frac{\dot{r}}{r}$$

and $\Gamma_{tt}^\varphi = \Gamma_{\varphi\varphi}^\varphi = \Gamma_{\varphi t}^t = \Gamma_{t\varphi}^t = 0$. List the covariant derivatives of the basis fields $\partial/\partial t$ and $\partial/\partial \varphi$ as vector fields along the meridians and latitudes. Determine the space of parallel vector fields along these curves. Discuss the special case of spheres.

3. If the first fundamental form of an immersion $f: M \longrightarrow \mathbb{R}^n$ is in diagonal form with respect to a chart (U, x) on M, that is, if $g_{ij} \equiv 0$ for $i \neq j$, then $g^{ii} = g_{ii}^{-1}$, $g^{ij} = 0$ for $i \neq j$ and therefore

$$\Gamma_{ik}^k = \frac{1}{2g_{kk}}\frac{\partial g_{kk}}{\partial x^i} \quad \text{and, for } i \neq k, \quad \Gamma_{ii}^k = \frac{-1}{2g_{kk}}\frac{\partial g_{ii}}{\partial x^k}.$$

∎

Exercise 4.6.15

Let $f: M \longrightarrow \mathbb{R}^3$ be an immersed surface together with a Gauß map $n: M \longrightarrow S^2$.

1. A smooth curve $c: I \longrightarrow M$ is a line of curvature if and only if there is a smooth function $\lambda: I \longrightarrow \mathbb{R}$ with $dn \circ \dot{c} = \lambda \cdot df \circ \dot{c}$.

2. (Joachimsthal's Theorem[18]) For $f_1: M_1 \longrightarrow \mathbb{R}^3$ and $f_2: M_2 \longrightarrow \mathbb{R}^3$, let c_1 in M_1 and c_2 in M_2 be curves with $f_1 \circ c_1 = f_2 \circ c_2 =: c$. Further let the intersection of f_1 and

---

[18]Ferdinand Joachimsthal (1818–1861).

f_2 along c be transversal, that is, $T_{c_1(t)}f_1 \neq T_{c_2(t)}f_2$ for all t. Show each two of the following statements implies the third:

(a) c_1 is a line of curvature;
(b) c_2 is a line of curvature;
(c) f_1 and f_2 intersect at a fixed angle along c.

3. We call $v \in T_pM$, $v \neq 0$, an *asymptotic direction* if $S_p^n(v, v) = 0$. Regular curves $c : I \longrightarrow M$, such that $\dot{c}(t)$ is an asymptotic direction for all $t \in I$, are called *asymptotes*. Check that the s-parameter lines on the ruled surface as in Example 4.3.12.3 are asymptotes. Note that $K(p) \leq 0$ if T_pM contains an asymptotic direction and that, up to collinearity, there are two asymptotic directions at p if $K(p) < 0$. ∎

Exercise 4.6.16

1. The *catenoid* is the surface of revolution generated by the catenary,

$$f = f(t, \varphi) = (\cosh t \cos \varphi, \cosh t \sin \varphi, t).$$

Compare the first fundamental form of the catenoid with the (somewhat differently parameterized than above) helicoid

$$f = f(t, \varphi) = (\sinh t \cos \varphi, \sinh t \sin \varphi, \varphi).$$

Determine the second fundamental forms of the catenoid and the helicoid and show that they are *minimal surfaces*, that is, that their mean curvatures H vanish. Compute their Gaußian curvatures. Sketch both surfaces.

2. Classify surfaces of revolution with constant Gaußian curvature and surfaces of revolution that are minimal surfaces, that is, have mean curvature $H \equiv 0$.

3. Let $f : M \longrightarrow \mathbb{R}^3$ be an immersed surface. For $x \in \mathbb{R}^3$, suppose the function $r : M \longrightarrow \mathbb{R}$, $r(p) = \|f(p) - x\|$ has a maximum at $p_0 \in M$. Show that $K(p_0) \geq 1/r(p_0)^2$ and conclude that M has a point with positive Gaußian curvature, if M is compact. (Compare with Exercise 4.6.3.3.) ∎

Exercise 4.6.17

Determine the curvature tensor of $G_k(V)$ and show that the sectional curvature of $G_k(V)$ is non-negative. ∎

Appendix A
Alternating Multilinear Forms

Werner Ballmann

© Springer Basel 2018
W. Ballmann, *Introduction to Geometry and Topology*, Compact Textbooks in Mathematics,
https://doi.org/10.1007/978-3-0348-0983-2

Let V be an n-dimensional vector space over a field K of characteristic 0. A map

$$T: V^k \longrightarrow K, \quad V^k := \underbrace{V \times \cdots \times V}_{k \text{ times}}, \tag{A.1}$$

is called k-*linear* or *multilinear*, if $T = T(v_1, \ldots, v_k)$ is linear in each of the variables v_i. We denote the vector space of k-linear maps $V^k \longrightarrow K$ by $L^k(V)$ and set $L^0(V) := K$.

For $S \in L^k(V)$ and $T \in L^l(V)$, we define $S \otimes T \in L^{k+l}(V)$ via

$$(S \otimes T)(v_1, \ldots, v_{k+l}) := S(v_1, \ldots, v_k) \cdot T(v_{k+1}, \ldots, v_{k+l}). \tag{A.2}$$

With this product, $\bigoplus_{k \geq 0} L^k(V)$ becomes an associative algebra. The unit of this multiplication is $1 \in K = L^0(V)$.

Let $L: W \longrightarrow V$ be linear. For $T \in L^k(V)$, we define $L^*T \in L^k(W)$ via

$$(L^*T)(w_1, \ldots, w_k) := T(Lw_1, \ldots, Lw_k). \tag{A.3}$$

The operation $T \mapsto L^*T$ is called *pullback* by L. Pullback by L is linear in L and T.

We call $T \in L^k(V)$ *alternating*, if

$$T(v_1, \ldots, v_i, \ldots, v_j, \ldots, v_k) = -T(v_1, \ldots, v_j, \ldots, v_i, \ldots, v_k) \tag{A.4}$$

for all $i < j$ and $v_1, \ldots, v_k \in V$. We denote the vector space of alternating $T \in L^k(V)$ by $A^k(V)$. We also call the elements of $A^k(V)$ *(alternating) k-forms*. We set $A^0(V) := L^0(V) = K$. For $L: W \longrightarrow V$ linear and $T \in A^k(V)$, we have $L^*T \in A^k(W)$.

For $T \in L^k(V)$, we define $\text{Alt } T \in L^k(V)$ by

$$\text{Alt } T(v_1, \ldots, v_k) := \frac{1}{k!} \sum_{\sigma \in S_k} \varepsilon(\sigma) \cdot T(v_{\sigma(1)}, \ldots, v_{\sigma(k)}), \tag{A.5}$$

where S_k denotes the symmetric group.

Lemma A.1 *For all $S \in L^k(V)$ and $T \in L^l(V)$ the following hold:*
1. $\text{Alt } S \in A^k(V)$;
2. $S \in A^k(V) \Longleftrightarrow \text{Alt } S = S$.
3. $\text{Alt}((\text{Alt } S) \otimes T) = \text{Alt}(S \otimes (\text{Alt } T)) = \text{Alt}(S \otimes T)$.

Together, the first two claims say that Alt is a projection from $L^k(V)$ onto $A^k(V)$.

Proof of Lemma A.1

We leave the proof of the first two claims as an exercise. To prove (3), let $G \cong S_k \subseteq S_{k+l}$ be the subgroup consisting of the $\sigma \in S_{k+l}$ with $\sigma(i) = i$, $k + 1 \leq i \leq k + l$. Then, for all $v_1, \ldots, v_{k+l} \in V$, the following holds:

$$\sum_{\sigma \in G} \varepsilon(\sigma) S(v_{\sigma(1)}, \ldots, v_{\sigma(k)}) \cdot T(v_{\sigma(k+1)}, \ldots, v_{\sigma(k+l)})$$

$$= \sum_{\sigma \in S_k} \varepsilon(\sigma) S(v_{\sigma(1)}, \ldots, v_{\sigma(k)}) \cdot T(v_{k+1}, \ldots, v_{k+l})$$

$$= k!((\text{Alt } S) \otimes T)(v_1, \ldots, v_{k+l}).$$

Now let τ be a representative of a coset $\tau G = \{\tau\sigma \mid \sigma \in G\}$ in S_{k+l} mod G. With $w_i := v_{\tau(i)}$, $1 \leq i \leq k + l$, it then holds that

$$\sum_{\sigma \in G} \varepsilon(\tau\sigma) S(v_{\tau\sigma(1)}, \ldots, v_{\tau\sigma(k)}) \cdot T(v_{\tau\sigma(k+1)}, \ldots, v_{\tau\sigma(k+l)})$$

$$= \varepsilon(\tau) \sum_{\sigma \in G} \varepsilon(\sigma) S(w_{\sigma(1)}, \ldots, w_{\sigma(k)}) \cdot T(w_{k+1}, \ldots, w_{k+l})$$

$$= \varepsilon(\tau) k!((\text{Alt } S) \otimes T)(w_1, \ldots, w_{k+l})$$

$$= \varepsilon(\tau) k!((\text{Alt } S) \otimes T)(v_{\tau(1)}, \ldots, v_{\tau(k+l)}).$$

It therefore follows that $\text{Alt}((\text{Alt } S) \otimes T) = \text{Alt}(S \otimes T)$, and the other equation follows analogously.   □

For $S \in A^k(V)$ and $T \in A^l(V)$ we define the *wedge product* $S \wedge T \in A^{k+l}(V)$ via

$$S \wedge T := \frac{(k+l)!}{k!l!} \text{Alt}(S \otimes T). \tag{A.6}$$

The prefactor $(k + l)!/k!l!$ is chosen precisely so that the computation rule (5) given below holds. In the literature one also finds other prefactors.

Computation Rule A.2 *The wedge product is*
1. *bilinear;*
2. *associative: for $R \in A^k(V)$, $S \in A^l(V)$ and $T \in A^m(V)$,*

$$(R \wedge S) \wedge T = \frac{(k+l+m)!}{k!l!m!} \text{Alt}(R \otimes S \otimes T) = R \wedge (S \wedge T);$$

3. *graded-commutative: for $S \in A^k(V)$ and $T \in A^l(V)$,*

$$S \wedge T = (-1)^{kl} T \wedge S;$$

4. *natural: if $L \colon W \longrightarrow V$ is linear, then*

$$L^*(S \wedge T) = L^* S \wedge L^* T;$$

and
5. *for $L^1, \ldots, L^k \in V^* = A^1(V)$ and $v_1, \ldots, v_k \in V$,*

$$\left(L^1 \wedge \cdots \wedge L^k\right)(v_1, \ldots, v_k) = \det\left(L^i\left(v_j\right)\right).$$

Proof
We leave the proofs of (1), (3) and (4) as exercises; (2) follows from Lemma A.1.3:

$$(R \wedge S) \wedge T = \frac{(k+l+m)!}{(k+l)!m!} \text{Alt}((R \wedge S) \otimes T)$$

$$= \frac{(k+l+m)!}{k!l!m!} \text{Alt}(\text{Alt}(R \otimes S) \otimes T)$$

$$= \frac{(k+l+m)!}{k!l!m!} \text{Alt}(R \otimes S \otimes T).$$

Claim (5) follows easily from (2). □

Corollary A.3 *Let $L^1, \ldots, L^k \in A^1(V)$. Then $L^1, \ldots, L^k$ are linearly independent if and only if $L^1 \wedge \cdots \wedge L^k \neq 0$.*

Proof
Let $L^1, \ldots, L^k$ be linearly independent. Then there is a basis $v_1, \ldots, v_n$ of V with $L^i(v_j) = \delta^i_j$, $1 \leq i \leq k$, $1 \leq j \leq n$. We therefore obtain

$$\left(L^1 \wedge \cdots \wedge L^k\right)(v_1, \ldots, v_k) = \det\left(\left(L^i\left(v_j\right)\right)\right)_{i,j} = 1 \neq 0.$$

If, on the other hand, there is a relation between the L^i, for example

$$L^1 = \alpha_2 L^2 + \cdots + \alpha_k L^k,$$

then

$$L^1 \wedge \cdots \wedge L^k = \left(\alpha_2 L^2 + \cdots + \alpha_k L^k\right) \wedge L^2 \wedge \cdots \wedge L^k$$

$$= \sum_{j=2}^{k} \alpha_j L^j \wedge L^2 \wedge \cdots \wedge L^k = 0,$$

since $\wedge$ is graded-commutative. $\square$

Corollary A.4 *Let* $(v_1, \ldots, v_n)$ *be a basis of* V *and* $(v^1, \ldots, v^n)$ *be the associated dual basis of* $V^* = A^1(V)$. *Then the tuple of*

$$v^{i_1} \wedge \cdots \wedge v^{i_k}, \quad 1 \le i_1 < i_2 < \cdots < i_k \le n, \tag{A.7}$$

is a basis of $A^k(V)$. *Therefore* $\dim A^k(V) = \binom{n}{k}$, *and, in particular,* $A^k(V) = \{0\}$ *for* $k > n$.

Proof
Clearly the $v^{i_1} \otimes \cdots \otimes v^{i_k}$, $1 \le i_1, \ldots, i_k \le n$, form a basis of $L^k(V)$. Under Alt, each of these is mapped to a multiple of $v^{i_1} \wedge \cdots \wedge v^{i_k}$, so the latter form a generating set for $A^k(V)$. Now let

$$\sum_{1 \le i_1 < \cdots < i_k \le n} \alpha_{i_1, \ldots, i_k} v^{i_1} \wedge \cdots \wedge v^{i_k}$$

be a linear combination of these, and let $1 \le j_1 < \ldots < j_k \le n$ be fixed. Then

$$(v^{i_1} \wedge \cdots \wedge v^{i_k})(v_{j_1}, \ldots, v_{j_k}) = \det((v^{i_\mu}(v_{j_\nu})))_{\mu,\nu} = \delta^{i_1}_{j_1} \cdots \delta^{i_k}_{j_k},$$

so

$$\sum_{i_1 < \cdots < i_k} (\alpha_{i_1, \ldots, i_k} v^{i_1} \wedge \cdots \wedge v^{i_k})(v_{j_1}, \ldots, v_{j_k}) = \alpha_{j_1, \ldots, j_k}.$$

Linear independence follows from this. $\square$

Lemma A.5 *Let* $v_1, \ldots, v_n$ *be a basis of* V, *and let* $w_1, \ldots, w_k \in V$. *Write* $w_i = \sum a_i^j v_j$, $1 \le i \le k$. *Then for* $T \in A^k(V)$,

$$T(w_1, \ldots, w_k) = \sum_{1 \le j_1 < \ldots < j_k \le m} \det\left(a_i^{j_\mu}\right) \cdot T\left(v_{j_1}, \ldots, v_{j_k}\right).$$

Proof

We compute

$$T\Big(\sum a_1^j v_j, \ldots, \sum a_k^j v_j\Big) = \sum_{1 \leq j_1, \ldots, j_k \leq m} a_1^{j_1} \cdots a_k^{j_k} \cdot T(v_{j_1}, \ldots, v_{j_k})$$

$$= \sum_{1 \leq j_1 < \ldots < j_k \leq m} \det\big(a_i^{j_\mu}\big) \cdot T(v_{j_1}, \ldots, v_{j_k}). \qquad \square$$

Corollary A.6 *Let* $\dim V = n$ *and* $T \in A^n(V) \setminus \{0\}$. *Then the condition* $T(v_1, \ldots, v_n) > 0$ *defines an orientation on* V. $\qquad \square$

Appendix B
Cochain Complexes

Werner Ballmann

© Springer Basel 2018

W. Ballmann, *Introduction to Geometry and Topology*, Compact Textbooks in Mathematics,
https://doi.org/10.1007/978-3-0348-0983-2

In the following R denotes a unital commutative ring. A reader unfamiliar with modules over rings may assume that R is a field, or, more particularly, the field $\mathbb{R}$ of real numbers. Modules over R are then R-vector spaces, and homomorphisms between them are R-linear maps.

Definition B.1

A *cochain complex* $\mathcal{C}$ over R consists of a sequence

$$\dots \xrightarrow{d} C^{k-1} \xrightarrow{d} C^k \xrightarrow{d} C^{k+1} \xrightarrow{d} \dots$$

of R-modules C^k, $k \in \mathbb{Z}$, and connecting homomorphisms, called *differentials* and here denoted somewhat laxly by d, such that at each step the composition vanishes, i.e. $d^2 = 0$.

Remark B.2 The reader may wonder if there might not also be chain complexes, since we are already talking about cochain complexes, and why we do not then discuss chain complexes before cochain complexes. In algebraic topology, chain complexes do, in fact, appear first, for example, the chain complex associated to a simplicial complex. On the level on which we will discuss cochain complexes in this appendix, the difference is of a purely formal nature: for a *chain complex*, the arrows point to the left, not to the right as in Definition B.1. We will not delve further into the topic at this point.

For a cochain complex $\mathcal{C}$ as in Definition B.1, we call the elements of C^k *cochains*, those of

$$Z^k = Z^k(\mathcal{C}) := \{ z \in C^k \mid dz = 0 \} \tag{B.1}$$

cocycles, and those of

$$B^k = B^k(\mathcal{C}) := d(C^{k-1}) \subseteq Z^k(\mathcal{C}) \tag{B.2}$$

coboundaries (in each case with the addition of "of degree k," if necessary). The *cohomology* of the C consists of the R-modules

$$H^k(C) := Z^k(C)/B^k(C). \tag{B.3}$$

We call the elements of $H^k(C)$ *cohomology classes* of C (of degree k). We call elements of Z^k which lie in the same cohomology class in $H^k(C)$ *cohomologous.*

Definition B.3

Let C_1 and C_2 be cochain complexes over R. A *homomorphism* $f: C_1 \longrightarrow C_2$ then consists of a sequence $f^k: C_1^k \longrightarrow C_2^k$ of homomorphisms, such that $f^{k+1}d_1 = d_2 f^k$ for all $k \in \mathbb{Z}$.

In the following, we will suppress the superscript (f^k) of morphisms, as we have already done for the differential. In this notation, the last condition in Definition B.3 says that the diagrams

$$
\begin{array}{ccc}
C_1^{k+1} & \xrightarrow{\ f\ } & C_2^{k+1} \\
{\scriptstyle d_1}\big\uparrow & & {\scriptstyle d_2}\big\uparrow \\
C_1^k & \xrightarrow{\ f\ } & C_2^k
\end{array}
\tag{B.4}
$$

are commutative for all $k \in \mathbb{Z}$.

Proposition B.4 *A homomorphism* $f: C_1 \longrightarrow C_2$ *of cochain complexes induces homomorphisms*

$$f^*: H^k(C_1) \longrightarrow H^k(C_2)$$

between their cohomologies. The identity on a cochain complex induces the identity on its cohomology, and the composition of homomorphisms of cochain complexes corresponds to the composition of induced homomorphisms, i.e. $(f \circ g)^* = f^* \circ g^*$.

Definition B.5

A *short exact sequence* of cochain complexes over R is a pair of homomorphisms

$$0 \longrightarrow C_1 \xrightarrow{\ i\ } C_2 \xrightarrow{\ j\ } C_3 \longrightarrow 0$$

among cochain complexes C_1, C_2 and C_3, such that i is injective, $\ker j = \operatorname{im} i$, and j is surjective.

In the formulation in Definition B.5, the zeros on the left and right play no role. They merely indicate that i is injective and j surjective, that is, that the kernel of i and the cokernel of j vanish: $\ker i = \{0\}$ and $\operatorname{coker} j = \{0\}$. By definition, a short exact sequence of cochain complexes as in Definition B.5 consists of an infinite commutative diagram of R-modules and homomorphisms, in which the columns are cochain complexes over R and in the rows each i is injective, $\ker j = \operatorname{im} i$, and each j is surjective:

A key phrase in the following is *diagram chase*. The meaning of this phrase will soon be made apparent to the reader. We consider a short exact sequence of cochain complexes as in Definition B.5. Then we obtain, for all $k \in \mathbb{Z}$, a homomorphism

$$\delta : H^k(\mathcal{C}_3) \longrightarrow H^{k+1}(\mathcal{C}_1). \tag{B.5}$$

To this end, let $c \in H^k(\mathcal{C}_3)$ and write $c = [z]$ with $z \in Z_3^k$, that is, $z \in C_3^k$ satisfies $d_3 z = 0$ and represents the equivalence class $c \in H^k(\mathcal{C}_3)$. Choose $y \in C_2^k$ with $jy = z$. This is possible, since j is surjective. We then have $jd_2 y = d_3 jy = d_3 z = 0$. Now $\ker j = \operatorname{im} i$, so there is an $x \in C_1^{k+1}$ with $ix = d_2 y$. This x is uniquely determined, since i is injective. Furthermore, $id_1 x = d_2 ix = d_2^2 y = 0$. Now, i is injective, so $d_1 x = 0$ and therefore $x \in Z_1^{k+1}$. We set

$$\delta c := [x] \in H^{k+1}(\mathcal{C}_1). \tag{B.6}$$

In the definition of δ, we made a choice in two places: We chose $z \in Z_3^k$ with $[z] = c$, and $y \in C_2^k$ with $jy = z$.

Lemma B.6 *The cohomology class $[x]$ of x does not depend on the choices of y and z. Therefore $\delta : H^k(\mathcal{C}_3) \longrightarrow H^{k+1}(\mathcal{C}_1)$ is a well-defined homomorphism of R-modules.*

Proof

Let $z' \in Z_3^k$ be another cocycle with $[z'] = c$. Then there is a $z'' \in C_3^{k-1}$ with $d_3 z'' = z' - z$. For z'', there is a $y'' \in C_2^{k-1}$ with $jy'' = z''$, and for this, we have $jd_2 y'' = d_3 jy'' = d_3 z'' = z' - z$. Therefore $j(y + d_2 y'') = z'$, so $y + d_2 y''$ is a permissible choice in place of y, and this choice satisfies $d_2(y + d_2 y'') = d_2 y$. We therefore obtain the same cocycle x as we did with the original choice of z.

Now let $y' \in C_2^k$ be another cochain with $jy' = z$. Then $j(y' - y) = jy' - jy = 0$, and so there exists $x' \in C_1^k$ with $ix' = y' - y$, hence $y' = y + ix'$. Therefore, $d_2 y' = d_2 y + d_2 ix' = d_2 y + id_1 x'$, and so $d_2 y' = i(x + d_1 x')$. The choice of y' instead of y therefore leads to the cohomology class $[x + d_1 x']$. Now, by definition, $[x] = [x + d_1 x']$, so δc is independent of the choice of y.   □

Proposition B.7 *For a short exact sequence of cochain complexes as in Definition B.5, the associated sequence*

$$\cdots \xrightarrow{\delta} H^k(\mathcal{C}_1) \xrightarrow{i^*} H^k(\mathcal{C}_2) \xrightarrow{j^*} H^k(\mathcal{C}_3) \xrightarrow{\delta} H^{k+1}(\mathcal{C}_1) \xrightarrow{i^*} \cdots,$$

is a long exact sequence, that is, at each term the image of the incoming homomorphism is equal to the kernel of the outgoing one.

Proof

We will work through one of the three necessary diagram chases and hope to thereby awake the thrill of the chase in the reader.

We follow the definition of δ above: Let $c \in H^k(\mathcal{C}_3)$ with $\delta c = 0$. Write $c = [z]$ for a cocycle $z \in C_3^k$ and choose $y \in C_2^k$ with $jy = z$. Then there is a unique cocycle $x \in C_1^{k+1}$ with $ix = d_2 y$. By definition $\delta c = [x]$. Since $[x] = \delta c = 0$, there is an $x' \in C_1^k$ with $d_1 x' = x$. Therefore $y' = y - ix' \in C_2^k$ is a cocycle,

$$d_2 y' = d_2 y - d_2 ix' = d_2 y - id_1 x' = d_2 y - ix = 0.$$

Additionally, $jy' = jy - jix' = jy = z$, so $j^*[y'] = [jy'] = [z] = c$, and thus $c \in \operatorname{im} j^*$. It therefore follows that $\ker \delta \subseteq \operatorname{im} j^*$. The other inclusion $\ker \delta \supseteq \operatorname{im} j^*$ is clear.   □

Bibliography

[AF] I. Agricola, T. Friedrich, *Global Analysis. Differential Forms in Analysis, Geometry and Physics*. Graduate Studies in Mathematics, vol. 52 (American Mathematical Society, Providence, 2008), xii+243 pp. Translated from the German by Andreas Nestke

[Bä] C. Bär, *Elementary Differential Geometry* (Cambridge University Press, Cambridge, 2010), 330 pp.

[BT] R. Bott, L. Tu, *Differential Forms in Algebraic Topology*. Graduate Texts in Mathematics, vol. 82 (Springer, Berlin, 1982), xiv+331 pp.

[BJ] T. Bröcker, K. Jänich, *Introduction to Differential Topology* (Cambridge University Press, Cambridge, 1973)

[Ch] S.S. Chern (ed.), *Global Differential Geometry*. MAA Studies in Mathematics, vol. 27. (Mathematical Association of America, Washington, 1989), ii+354 pp.

[CR] R. Courant, H. Robbins, *What is Mathematics?* 2nd edn. (Oxford University Press, Oxford, 1999), 592 pp. Revised by Ian Stewart

[dC] M. do Carmo. *Differential Geometry of Curves and Surfaces* (Prentice-Hall, Princeton, 1976), 503 pp.

[EJ] J.-H. Eschenburg, J. Jost, *Differentialgeometrie und Minimalflächen*, Zweite Auflage. Graduate Texts in Mathematics, vol. 82 (Springer, Berlin, 2007), xvi+256 pp.

[Ha] A. Hatcher, *Algebraic Topology* (Cambridge University Press, Cambridge, 2001), xii+544 pp.

[HT] S. Hildebrandt, A. Tromba, *The Parsimonious Universe. Shape and Form in the Natural World* (Copernicus, New York, 1996)

[Ho] H. Hopf, *Differential Geometry in the Large*, 2nd edn. With a preface by K. Voss. Lecture Notes in Mathematics, vol. 1000 (Springer, Berlin, 1989), viii+184 pp. Notes taken by Peter Lax and John W. Gray. With a preface by S. S. Chern

[Kel] J.L. Kelley, *General Topology*. Graduate Texts in Mathematics, vol. 27 (Springer, New York, 1975), xiv+298 pp.

[Ke] M. Kervaire, A manifold which does not admit any differentiable structure. Comment. Math. Helv. **34**, 257–270 (1960)

[Kl] W. Klingenberg, *A Course in Differential Geometry*. Graduate Texts in Mathematics, vol. 51 (Springer, Berlin, 1978), xii+178 pp.

[Kü] W. Kühnel, *Differential Geometry: Curves – Surfaces – Manifolds*, 2nd edn. (American Mathematical Society, Providence, 2005), 380 pp.

[La] S. Lang, *Real Analysis* (Addison-Wesley, Reading, 1969), xi+476 pp.

[Le] J.M. Lee, *Introduction to Smooth Manifolds*. Graduate Texts in Mathematics (Springer, Berlin, 2003), xvii+628 pp.

[MP] R.S. Millman, G.D. Parker, *Elements of Differential Geometry* (Prentice-Hall, Princeton, 1977), xiv+265 pp.

[Mi1] J. Milnor, *Morse Theory*. Annals of Mathematics Studies, vol. 51 (Princeton University Press, Princeton, 1963), vi+153 pp. Based on lecture notes by M. Spivak and R. Wells

[Mi2] J. Milnor, *Lectures on the h-Cobordism Theorem* (Princeton University Press, Princeton, 1965), v+116 pp. Notes by L. Siebenmann and J. Sondow

[Mi3] J. Milnor, *Topology from the Differentiable Viewpoint* (The University Press of Virginia, Charlottesville, 1965), ix+65 pp. Based on notes by David W. Weaver

[ST] I.M. Singer, J.A. Thorpe, *Lecture Notes on Elementary Topology and Geometry,* Reprint of the 1967 edition. Undergraduate Texts in Mathematics (Springer, Berlin, 1976), viii+232 pp.

© Springer Basel 2018

W. Ballmann, *Introduction to Geometry and Topology*, Compact Textbooks in Mathematics, https://doi.org/10.1007/978-3-0348-0983-2

[Sp1] M. Spivak, *A Comprehensive Introduction to Differential Geometry. Vol. I*, 2nd edn. (Publish or Perish, Berkeley, 1979), xiv+668 pp.

[Sp2] M. Spivak, *A Comprehensive Introduction to Differential Geometry. Vol. II*, 2nd edn. (Publish or Perish, Berkeley, 1979), xv+423 pp.

[Sp3] M. Spivak, *A Comprehensive Introduction to Differential Geometry. Vol. III*, 2nd edn. (Publish or Perish, Berkeley, 1979), xii+466 pp.

[St1] K. Strubecker, *Differentialgeometrie. I. Kurventheorie der Ebene und des Raumes*. Sammlung Göschen Bd. 1113/1113a. (de Gruyter, Berlin, 1955), 150 pp.

[St2] K. Strubecker, *Differentialgeometrie. II. Theorie der Flächenmetrik*. Sammlung Göschen Bd. 1179/1179a (de Gruyter, Berlin, 1958), 195 pp.

[St3] K. Strubecker, *Differentialgeometrie. III. Theorie der Flächenkrümmung*. Sammlung Göschen Bd. 1180/1180a (de Gruyter, Berlin, 1959), 254 pp.

[Wh1] H. Whitney, Differentiable manifolds. Ann. Math. **37**, 645–680 (1936)

[Wh2] H. Whitney, *Geometric Integration Theory* (Princeton University Press, Princeton, 1957), xv+387 pp.

Index

© Springer Basel 2018
W. Ballmann, *Introduction to Geometry and Topology*, Compact Textbooks in Mathematics,
https://doi.org/10.1007/978-3-0348-0983-2

Printed in the United States
by Bookmasters